Brooklands Books

CW01583074

SUBARU IMPREZA TURBO

1994 2001

Limited Edition Extra

Compiled by
R.M.Clarke

ISBN 1 85520 5696

BROOKLANDS BOOKS LTD.
P.O. BOX 146, COBHAM,
SURREY, KT11 1LG. UK
sales@brooklands-books.com

A-SUITX2

Printed in China

Brooklands Books

MOTORING

B.B. ROAD TEST SERIES

Abarth Gold Portfolio 1950-1971
AC Ace & Aceca 1953-1983
Alfa Romeo Giulietta Gold Portfolio 1954-1965
Alfa Romeo Giulia Coupés 1963-1976
Alfa Romeo Giulia Coupés Gold Port. 1963-1976
Alfa Romeo Spider 1966-1990
Alfa Romeo Spider Gold Portfolio 1966-1991
Alfa Romeo Alfasud 1972-1984
Alfa Romeo Alfetta Gold Portfolio 1972-1987
Alfa Romeo Alfetta GTV6 1980-1986
Alvis Gold Portfolio 1919-1967
AMX & Javelin Muscle Portfolio 1968-1974
Armstrong Siddeley Gold Portfolio 1945-1960
Aston Martin Gold Portfolio 1948-1971
Aston Martin Gold Portfolio 1972-1985
Aston Martin Gold Portfolio 1985-1995
Audi Quattro Gold Portfolio 1980-1991
Audi Quattro Takes On The Competition
Austin A30 & A35 1951-1962
Austin-Healey 100 & 100/6 Gold Port. 1952-1959
Austin-Healey 3000 Ultimate Portfolio 1959-1967
Austin-Healey Sprite Gold Portfolio 1958-1971
Berkeley Sportscars Limited Edition
BMW 6 & 8 Cyl. Cars Limited Edition 1935-1960
BMW 1600 Collection No. 1 1966-1981
BMW 2002 Gold Portfolio 1968-1976
BMW 6 Cylinder Coupés & Saloons Gold P. 1969-1976
BMW 316, 318, 320 (4 cyl.) Gold Port. 1975-1990
BMW 320, 323, 325 (6 cyl.) Gold Port. 1977-1990
BMW 3 Series Gold Portfolio 1991-1997
BMW 5 Series Gold Portfolio 1981-1987
BMW 5 Series Gold Portfolio 1988-1995
BMW 6 Series Gold Portfolio 1976-1989
BMW 7 Series Performance Portfolio 1977-1986
BMW 8 Series Limited Edition
BMW Alpina Performance Portfolio 1967-1987
BMW Alpina Performance Portfolio 1988-1998
BMW M Series Gold Portfolio 1976-1997
BMW Z3 & Z3M Limited Edition
Borgward Isabella Limited Edition
Bricklin Gold Portfolio 1974-1975
Bristol Cars Portfolio
Buick Performance Portfolio 1947-1962
Buick Muscle Portfolio 1963-1973
Buick Riviera Performance Portfolio 1963-1978
Cadillac Allanté 1986-1993
Cadillac Automobiles 1949-1959
Cadillac Automobiles 1960-1969
Cadillac Eldorado Performance Portfolio 1967-1978
Checker Limited Edition
Chevrolet 1955-1957
Impala & SS Muscle Portfolio 1958-1972
Corvair Performance Portfolio 1959-1969
El Camino & SS Muscle Portfolio 1959-1987
Chevy II & Nova SS Muscle Portfolio 1962-1974
Chevelle & SS Muscle Portfolio 1964-1972
Caprice Limited Edition 1965-1976
Chevrolet Muscle Cars 1966-1971
Chevy Blazer 1969-1981
Camaro Muscle Portfolio 1967-1973
High Performance Camaros 1982-1988
Camaro Performance Portfolio 1993-2000
Chevrolet Corvette Gold Portfolio 1953-1962
Chevrolet Corvette Sting Ray Gold Port. 1963-1967
Chevrolet Corvette Gold Portfolio 1968-1977
High Performance Corvettes 1983-1989
Chrysler 300 Gold Portfolio 1955-1970
Valiant 1960-1962
Citroen Traction Avant Gold Portfolio 1934-1957
Citroen 2CV Ultimate Portfolio 1948-1990
Citroen DS & ID 1955-1975
Citroen DS & ID Gold Portfolio 1955-1975
Citroen SM Limited Edition Extra 1970-1975
Shelby Cobra Gold Portfolio 1962-1969
Cobras & Cobra Replicas Gold Portfolio 1962-1989
Crosley & Crosley Specials Limited Edition
Cunningham Automobiles 1951-1955
Datsun Roadsters Performance Portfolio 1960-71
Datsun 240Z & 260Z Gold Portfolio 1970-1978
Datsun 280Z & ZX 1975-1983
DeLorean Gold Portfolio 1977-1995
De Soto Limited Edition 1952-1960
Charger Muscle Portfolio 1966-1974
Dodge Viper Performance Portfolio 1990-1998
ERA Gold Portfolio 1934-1994
Facel Vega 1954-1964
Ferrari Limited Edition 1947-1957
Ferrari Limited Edition 1958-1963
Ferrari Dino Limited Edition Extra 1965-1974
Ferrari Dino 308 & Mondial Gold Portfolio 1974-1985
Ferrari 328 348 Mondial Ultimate Portfolio 1986-94
Fiat 500 Gold Portfolio 1936-1972
Fiat 600 & 850 Gold Portfolio 1955-1972
Fiat Pininfarina 124 & 2000 Spider 1968-1985
Fiat X1/9 Gold Portfolio 1973-1989
Fiat Abarth Performance Portfolio 1972-1987
Ford Consul, Zephyr, Zodiac Mk. I & II 1950-1962
Ford Zephyr, Zodiac, Executive Mk. III & IV 1962-1971
Ford Cortina 1600E & GT 1967-1970
High Performance Capris Gold Portfolio 1969-1987
Capri Muscle Portfolio 1974-1987
High Performance Fiestas 1979-1991
Ford Escort RS & Mexico Limited Edition 1970-1979
High Performance Escorts Mk. I 1968-1974
High Performance Escorts Mk. II 1975-1980
High Performance Escorts 1980-1985
High Performance Escorts 1985-1990
High Perf. Sierras & Merkurs Gold Port. 1983-1990
Ford Thunderbird Performance Portfolio 1955-1957
Ford Thunderbird Performance Portfolio 1958-1963
Ford Thunderbird Performance Portfolio 1964-1976
Ford Automobiles 1949-1959
Ford Fairlane Performance Portfolio 1955-1970
Ford Ranchero Muscle Portfolio 1957-1979
Edsel Limited Edition 1957-1960
Falcon Performance Portfolio 1960-1970
Ford Galaxie & LTD Limited Edition 1960-1973
Ford GT40 Gold Portfolio 1964-1987
Ford Torino Limited Edition 1968-1974
Ford Bronco 4x4 Performance Portfolio 1966-1977
Ford Bronco 1978-1988
Goggomobil Limited Edition
Holden 1948-1962
Honda S500 • S600 • S800 Limited Edition 1962-1970
Honda CRX 1983-1987
International Scout Gold Portfolio 1961-1980
Isetta Gold Portfolio 1953-1964
ISO & Bizzarrini Gold Portfolio 1962-1974
Jaguar and SS Gold Portfolio 1931-1951
Jaguar C-Type & D-Type Gold Portfolio 1951-1960
Jaguar XK120, 140, 150 Gold Portfolio 1948-1960
Jaguar Mk. VII, VIII, IX, X, 420 Gold Port. 1950-1970
Jaguar Mk. 1 & Mk. 2 Gold Portfolio 1955-1969
Jaguar E-Type Gold Portfolio 1961-1971
Jaguar E-Type V-12 1971-1975
Jaguar S-Type & 420 Limited Edition 1963-1968
Jaguar XJ12, XJ5.3, V12 Gold Portfolio 1972-1990
Jaguar XJ6 Series I & II Gold Portfolio 1968-1979
Jaguar XJ6 Series III Perf. Portfolio 1979-1986
Jaguar XJ6 Gold Portfolio 1986-1994
Jaguar XJS Gold Portfolio 1975-1988
Jaguar XJ-S V12 Ultimate Portfolio 1988-1996
Jaguar XK8 Limited Edition
Jeep CJ-5 & CJ-6 1960-1976
Jeep CJ-5 & CJ-7 4x4 Perf. Portfolio 1976-1986
Jeep Wagoneer Performance Portfolio 1963-1991
Jeep J-Series Pickups 1970-1982
Jeepster & Commando Limited Edition 1967-1973
Jeep Cherokee & Comanche Pickups P. P. 1984-91
Jeep Wrangler 4x4 Performance Portfolio 1987-99
Jeep Cherokee & Grand Cherokee 4x4 P. P. 1992-98
Jensen Interceptor Gold Portfolio 1966-1986
Jensen - Healey Limited Edition 1972-1976
Kaiser - Frazer Limited Edition 1946-1955
Lagonda Gold Portfolio 1919-1964
Lancia Aurelia & Flaminia Gold Portfolio 1950-1970
Lancia Fulvia Gold Portfolio 1963-1976
Lancia Beta Gold Portfolio 1972-1984
Lancia Stratos 1972-1985
Lancia Delta & integrale Ultimate Portfolio
Land Rover Series I 1948-1958
Land Rover Series II & IIa 1958-1971
Land Rover Series III 4x4 Perf. Portfolio 1971-1985
Land Rover 90 110 Defender Gold Portfolio 1983-1994
Land Rover Discovery Perf. Port. 1989-2000
Land Rover Story Part One 1948-1971
Fifty Years of Selling Land Rover
Lincoln Gold Portfolio 1949-1960
Lincoln Continental Performance Portfolio 1961-1969
Lincoln Continental 1969-1976
Lotus Sports Racers Portfolio - covering 1951-1965
Lotus Seven Gold Portfolio 1957-1973
Lotus Elite Limited Edition 1957-1964
Lotus Elan Ultimate Portfolio 1962-1974
Lotus Elan & SE 1989-1992
Lotus Europa Gold Portfolio 1966-1975
Lotus Elite & Eclat 1974-1982
Lotus Elise Limited Edition
Marcos Coupés & Spyders Gold Portfolio 1960-1997
Matra Limited Edition 1965-1983
Mazda Miata MX-5 Performance Portfolio 1989-1997
Mazda Miata MX-5 Takes On The Competition
Mazda RX-7 Gold Portfolio 1978-1991
McLaren F1 Sportscar Limited Edition
Mercedes 190 & 300 SL 1954-1963
Mercedes G-Wagen 1981-1994
Mercedes S & 600 1965-1972
Mercedes S Class 1972-1979
Mercedes S Class Limited Edition Extra 1980-1991
Mercedes 230 • 250 • 280SL Gold Portfolio 1963-1971
Mercedes SLs & SLCs Gold Portfolio 1971-1989
Mercedes SLs Performance Portfolio 1989-1994
Mercedes 190 Limited Edition Extra 1983-1993
Mercedes CLK & SLK Limited Edition
Mercury Limited Edition 1947-1959
Mercury Comet & Cyclone Limited Edition 1960-1970
Cougar Muscle Portfolio 1967-1973
Messerschmitt Gold Portfolio 1954-1964
MG Gold Portfolio 1929-1939
MG TA & TC Gold Portfolio 1936-1949
MG TD & TF Gold Portfolio 1949-1955
MGA & Twin Cam Gold Portfolio 1955-1962
MG Midget Gold Portfolio 1961-1979
MGB Roadsters 1962-1980
MGB MGC & V8 Gold Portfolio 1962-1980
MGB GT 1965-1980
MGC & MGB GT V8 Limited Edition
MG Y-Type & Magnette ZA/ZB Limited Edition
MGF Limited Edition
Mini Gold Portfolio 1959-1969
Mini Gold Portfolio 1969-1980
Mini Gold Portfolio 1981-1997
High Performance Minis Gold Portfolio 1960-1973
Mini Cooper Gold Portfolio 1961-1971
Mini Moke Gold Portfolio 1964-1994
Morgan Three-Wheeler Gold Portfolio 1910-1952
Morgan Plus 4 & Four 4 Gold Portfolio 1936-1967
Morgan Cars Portfolio 1968-2001
Morris Minor Collection No. 1 1948-1980
Shelby Mustang Muscle Portfolio 1965-1970
Mustang Muscle Portfolio 1967-1973
High Performance Mustang IIs 1974-1978
Mustang 5.0L Muscle Portfolio 1982-1993
Mustang 5.0L Takes On The Competition
Nash & Nash-Healey Limited Edition 1949-1957
Nash-Austin Metropolitan Gold Portfolio 1954-1962
NSU Ro80 Limited Edition
NSX Performance Portfolio 1989-1999
Oldsmobile Automobiles 1955-1963
Oldsmobile Muscle Portfolio 1964-1971
Cutlass & 4-4-2 Muscle Portfolio 1964-1974
Oldsmobile Toronado 1966-1978
Opel GT Gold Portfolio 1968-1973
Opel Manta Limited Edition 1970-1975
Packard Gold Portfolio 1946-1958
Pantera Ultimate Portfolio 1970-1995
Panther Gold Portfolio 1972-1990
Barracuda Muscle Portfolio 1964-1974
Pontiac Limited Edition 1949-1960
Pontiac Tempest & GTO 1961-1965
GTO Muscle Portfolio 1964-1974
Firebird & Trans-Am Muscle Portfolio 1967-1972
Firebird & Trans-Am Muscle Portfolio 1973-1981
High Performance Firebirds 1982-1988
Firebird & Trans Am Performance Portfolio 1993-2000
Pontiac Fiero Performance Portfolio 1984-1988
Porsche 356 Gold Portfolio 1953-1965
Porsche 912 Limited Edition
Porsche 911 1965-1969
Porsche 911 1970-1972
Porsche 911 1973-1977
Porsche 911 SC & Turbo Gold Portfolio 1978-1983
Porsche 911 Carrera & Turbo Gold Port. 1984-1989
Porsche 911 Gold Portfolio 1990-1997
Porsche 911 Takes On The Competition 1990-1997
Porsche 914 Ultimate Portfolio
Porsche 924 Gold Portfolio 1975-1988
Porsche 928 Performance Portfolio 1977-1994
Porsche 928 Takes On The Competition
Porsche 944 Ultimate Portfolio
Porsche 968 Limited Edition
Porsche Boxster Limited Edition
Railton & Brough Superior Gold Portfolio 1933-1950
Range Rover Gold Portfolio 1970-1985
Range Rover Gold Portfolio 1985-1995
Range Rover Takes on the Competition
Reliant Scimitar 1964-1986
Renault Alpine Gold Portfolio 1958-1994
Riley Gold Portfolio 1924-1939
R. R. Silver Cloud & Bentley 'S' Series Gold P. 1955-65
Rolls Royce Silver Shadow Ultimate Portfolio 1965-80
Rolls Royce & Bentley Gold Portfolio 1980-1989
Rolls Royce & Bentley Limited Edition 1990-1997
Rover P4 1949-1959
Rover 3 & 3.5 Litre Gold Portfolio 1958-1973
Rover 2000 & 2200 1963-1977
Saab Sonett Collection No.1 1966-1974
Saab Turbo 1976-1983
Studebaker Gold Portfolio 1947-1966
Studebaker Hawks & Larks 1956-1963
Avanti 1962-1990
Starion & Conquest Performance Portfolio 1982-90
Subaru Impreza Turbo Limited Edition Extra 1994-2001
Suzuki SJ Gold Portfolio 1971-1997
Vitara, Sidekick & Geo Tracker Perf. Port. 1988-1997
Sunbeam Tiger & Alpine Gold Portfolio 1959-1967
Toyota Land Cruiser Gold Portfolio 1956-1987
Toyota Land Cruiser 1988-1997
Toyota MR2 Gold Portfolio 1984-1997
Toyota MR2 Takes On The Competition
Triumph TR2 & TR3 Gold Portfolio 1952-1961
Triumph TR4, TR5, TR250 1961-1968
Triumph TR6 Gold Portfolio 1969-1976
Triumph Herald 1959-1971
Triumph Vitesse 1962-1971
Triumph Spitfire Gold Portfolio 1962-1980
Triumph 2000, 2.5, 2500 1963-1977
Triumph GT6 Gold Portfolio 1966-1974
Triumph Stag Gold Portfolio 1970-1977
Triumph Dolomite Sprint Limited Edition
TVR Gold Portfolio 1959-1986
TVR Performance Portfolio 1986-1994
TVR Performance Portfolio 1995-2000
VW Beetle Gold Portfolio 1935-1967
VW Beetle Gold Portfolio 1968-1991
VW Beetle Collection No.1 1970-1982
VW Karmann Ghia 1955-1982
VW Bus, Camper, Van 1954-1967
VW Bus, Camper, Van Perf. Portfolio 1968-1979
VW Bus, Camper, Van 1979-1989
VW Scirocco 1974-1981
Volvo PV444 & PV544 Perf. Portfolio 1945-1965
Volvo 120 Amazon Ultimate Portfolio
Volvo 1800 Gold Portfolio 1960-1973
Volvo 140 & 160 Series Gold Portfolio 1966-1975
Forty Years of Selling Volvo
Westfield Limited Edition

B.B. ROAD & TRACK SERIES

Road & Track on Alfa Romeo 1964-1970
Road & Track on Alfa Romeo 1971-1976
Road & Track on Aston Martin 1962-1990
Road & Track on Audi & Auto Union 1952-1980
Road & Track on Audi & Auto Union 1980-1986
Road & Track on Austin Healey 1953-1970
Road & Track on BMW Cars 1966-1974
Road & Track on BMW Cars 1975-1978
Road & Track on BMW Cars 1979-1983
R & T on Cobra, Shelby & Ford GT40 1962-1992
Road & Track on Corvette 1953-1967
Road & Track on Corvette 1968-1982
Road & Track on Corvette 1982-1986
Road & Track on Corvette 1986-1990
Road & Track on Ferrari 1975-1981
Road & Track on Ferrari 1981-1984
Road & Track on Ferrari 1984-1988
Road & Track on Fiat Sports Cars 1968-1987
Road & Track on Jaguar 1950-1960
Road & Track on Jaguar 1961-1968
Road & Track on Jaguar 1968-1974
Road & Track on Jaguar 1974-1982
Road & Track on Jaguar 1983-1989
Road & Track on Lamborghini 1964-1985
R & T on Mazda RX-7 & MX-5 Miata 1986-1991
Road & Track on Mercedes 1952-1962
Road & Track on Mercedes 1963-1970
Road & Track on Mercedes 1971-1979
Road & Track on Mercedes 1980-1987
Road & Track on MG Sports Cars 1949-1961
Road & Track on MG Sports Cars 1962-1980
R & T on Nissan 300-ZX & Turbo 1984-1989
Road & Track on Pontiac 1960-1983
Road & Track on Porsche 1951-1967
Road & Track on Porsche 1968-1971
Road & Track on Porsche 1972-1975
Road & Track on Porsche 1975-1978
Road & Track on Porsche 1979-1982
Road & Track on Porsche 1985-1988
R & T on Rolls Royce & Bentley 1950-1965
R & T on Rolls Royce & Bentley 1966-1984
Road & Track on Saab 1972-1992
R & T on Toyota Sports & GT Cars 1966-1984
R & T on Triumph Sports Cars 1953-1967
R & T on Triumph Sports Cars 1967-1974
R & T on Triumph Sports Cars 1974-1982
Road & Track on Volkswagen 1951-1968
Road & Track on Volkswagen 1968-1978
Road & Track on Volkswagen 1978-1985
Road & Track on Volvo 1957-1974
Road & Track on Volvo 1977-1994
Road & Track - Henry Manney at Large & Abroad
Road & Track - Best of PS
Road & Track - Peter Egan "At Large"
Road & Track - Peter Egan Side Glances 1983-92
Road & Track - Peter Egan Side Glances 1992-97

B.B. CAR AND DRIVER SERIES

Car and Driver on BMW 1955-1977
Car and Driver on Corvette 1978-1982
Car and Driver on Corvette 1983-1988
C and D on Datsun Z 1600 & 2000 1966-1984
Car and Driver on Ferrari 1955-1962
Car and Driver on Ferrari 1963-1975
Car and Driver on Ferrari 1976-1983
Car and Driver on Mopar 1956-1967
Car and Driver on Mustang 1964-1972
Car and Driver on Pontiac 1961-1975
Car and Driver on Porsche 1955-1962
Car and Driver on Porsche 1963-1970
Car and Driver on Porsche 1970-1976
Car and Driver on Porsche 1977-1981
Car and Driver on Porsche 1982-1986
Car and Driver on Volvo 1955-1986

RACING & THE LAND SPEED RECORD

The Land Speed Record 1898-1919
The Land Speed Record 1920-1929
The Land Speed Record 1930-1939
The Land Speed Record 1940-1962
The Land Speed Record 1963-1999
The Land Speed Record 1898-1999 - Hard Bound
Can-Am Racing 1966-1969
Can-Am Racing 1970-1974
Can-Am Racing Cars 1966-1974
The Carrera Panamericana Mexico - 1950-1954
Le Mans - The Bentley & Alfa Years - 1923-1939
Le Mans - The Jaguar Years - 1949-1957
Le Mans - The Ferrari Years - 1958-1965
Le Mans - The Ford & Matra Years - 1966-1974
Le Mans - The Porszche Years - 1975-1982
Le Mans - The Porsche & Jaguar Years - 1983-91
Le Mans - The Porsche & Peugeot Years - 1992-99
Le Mans - 1923-1999 - Hard Bound
Mille Miglia - The Alfa & Ferrari Years - 1927-1951
Mille Miglia - The Ferrari & Mercedes Years - 1952-57
Targa Florio - The Post War Years - 1948-1973 - H.B.
Targa Florio - The Porsche & Ferrari Years - 1955-1964
Targa Florio - The Porsche Years - 1965-1973

B.B. PRACTICAL CLASSICS SERIES

PC on Austin A40 Restoration
PC on Land Rover Restoration
PC on Midget/Sprite Restoration
PC on MGB Restoration
PC on Sunbeam Rapier Restoration
PC on Triumph Herald/Vitesse

B.B. HOT ROD 'ENGINE' SERIES

Chevy 265 & 283
Chevy 302 & 327
Chevy 348 & 409
Chevy 396 & 427
Chevy 454 thru 512
Chrysler Hemi
Chrysler 273, 318, 340 & 360
Chrysler 361, 383, 400, 413, 426 & 440
Ford 289, 302, Boss 302 & 351W
Ford 351C & Boss 351

B.B. RESTORATION & GUIDE SERIES

BMW 2002 - A Comprehensive Guide
BMW '02 Restoration Guide
Classic Camaro Restoration
Chevrolet High Performance Tips & Techniques
Chevy Engine Swapping Tips & Techniques
Chevy-GMC Pickup Repair
Engine Swapping Tips & Techniques
Land Rover Restoration Portfolio
MG 'T' Series Restoration Guide
MGA Restoration Guide
Mustang Restoration Tips & Techniques
Practical Gas Flow
The Great Classic Muscle Cars Compared

MOTORCYCLING

B.B. ROAD TEST SERIES

AJS & Matchless Gold Portfolio 1945-1966
BMW Motorcycles Gold Portfolio 1950-1971
BMW Motorcycles Gold Portfolio 1971-1976
BSA Singles Gold Portfolio 1945-1963
BSA Singles Gold Portfolio 1964-1974
BSA Twins A7 & A10 Gold Portfolio 1946-1962
BSA Twins A50 & A65 Gold Portfolio 1962-1973
BSA & Triumph Triples Gold Portfolio 1968-1976
Ducati Gold Portfolio 1974-1978
Ducati Gold Portfolio 1978-1982
Harley-Davidson Sportsters Perf. Port. 1965-1976
Harley-Davidson Super Glide Perf. Port. 1971-1981
Harley-Davidson FXR Series Perf. Port. 1982-1992
Honda CB750 Gold Portfolio 1969-1978
Honda CB500 & 550 Fours Perf. Port. 1971-1977
Honda CB350 & 400 Fours Perf. Port. 1972-1978
Honda Gold Wing Gold Portfolio 1975-1995
Honda CBX 1000 Gold Portfolio 1978-1982
Honda RC30 Performance Portfolio 1988-1992
Kawasaki Z1 900 Performance Portfolio 1972-1977
Kawasaki 500 & 750 Triples Perf. Port. 1969-1976
Kawasaki GPZ 900R Ninja Perf. Port. 1984-1996
Laverda Gold Portfolio 1967-1977
Laverda Performance Portfolio 1978-1988
Laverda Jota Performance Portfolio 1976-1985
Moto Guzzi Gold Portfolio 1949-1973
Moto Guzzi Le Mans Performance Portfolio 1976-89
Moto Morini 3½ & 500 Performance Port. 1974-84
Norton Dominators Performance Portfolio 1949-70
Norton Commando Ultimate Portfolio 1968-1977
Norton Rotaries Limited Edition Extra 1984-1992
Suzuki GT750 Performance Portfolio 1971-1977
Suzuki GS1000 Performance Portfolio 1978-1981
Vincent Gold Portfolio 1945-1980
Yamaha RD350/400 Performance Portfolio 1972-79
Yamaha XS650 Performance Portfolio 1969-1985

B.B. CYCLE WORLD SERIES

Cycle World on BMW 1974-1980
Cycle World on BMW 1981-1986
Cycle World on Ducati 1982-1991
Cycle World on Harley-Davidson 1978-1983
Cycle World on Harley-Davidson 1983-1987
Cycle World on Harley-Davidson 1987-1990
Cycle World on Harley-Davidson 1990-1992
Cycle World on Honda 1962-1967
Cycle World on Honda 1968-1971
Cycle World on Honda 1971-1974
Cycle World on Husqvarna 1966-1976
Cycle World on Husqvarna 1977-1984
Cycle World on Kawasaki 1966-1971
Cycle World on Kawasaki Off-Road Bikes 1972-1979
Cycle World on Kawasaki Street Bikes 1972-1976
Cycle World on Norton 1962-1971
Cycle World on Suzuki 1962-1970
Cycle World on Suzuki Off-Road Bikes 1971-1976
Cycle World on Suzuki Street Bikes 1971-1976
Cycle World on Triumph 1967-1972
Cycle World on Yamaha 1962-1969
Cycle World on Yamaha Off-Road Bikes 1970-1974
Cycle World on Yamaha Street Bikes 1970-1974

MILITARY

VEHICLES & AEROPLANES

Complete WW2 Military Jeep Manual
Dodge WW2 Military Portfolio 1940-1945
German Military Equipment WW2
Hail To The Jeep
Land Rover Military Portfolio
Military & Civilian Amphibians 1940-1990
Off Road Jeeps Civilian & Military 1944-1971
Silhouette Handbook of US Army Air Forces Aeroplanes
US Military Vehicles 1941-1945
US Army Military Vehicles WW2-TM9-2800
VW Kubelwagen Military Portfolio 1940-1990
WW2 Allied Vehicles Military Portfolio 1939-1945
WW2 Jeep Military Portfolio 1941-1945

10/6Z1

Brooklands Books

CONTENTS

ACKNOWLEDGEMENTS

For more than 40 years, Brooklands Books have been publishing compilations of road tests and other articles from the English speaking world's leading motoring magazines. We have already published more than 700 titles, and in these we have made available to motoring enthusiasts some 25,000 stories which would otherwise have become hard to find. For the most part, our books focus on a single model, and as such they have become an invaluable source of information. As Bill Boddy of *Motor Sport* was kind enough to write when reviewing one of our Gold Portfolio volumes, the Brooklands catalogue "must now constitute the most complete historical source of reference available, at least of the more recent makes and models."

Even so, we are constantly being asked to publish new titles on cars which have a narrower appeal than those we have already covered in our main series. The economics of book production make it impossible to cover these subjects in our main series, but Limited Edition and Limited Edition Extra volumes, like this one, give us a way to tackle these less popular but no less worthy subjects.

Both the Limited Edition and Limited Edition Extra series maintain the same high standards of presentation and reproduction set by our established ranges. However, each volume is printed in smaller quantities - which is perhaps the best reason we can think of why you should buy this book now. We would also like to remind readers that we are always open to suggestions for new titles; perhaps your club or interest group would like us to consider a book on your particular subject?

Finally, we are more than pleased to acknowledge that Brooklands Books rely on the help and co-operation of those who publish the magazines where the articles in our books originally appeared. For this present volume, we gratefully acknowledge the continued support of the publishers of *Autocar, Autosport, Car, Car and Driver, Car Australia, Car South Africa, Cars & Car Conversions, Motor Sport, NZ Classic Car, Open Road, Performance Car, Top Gear, What Car?* and *Wheels* for allowing us to include their valuable and informative copyright stories.

R.M. Clarke.

FIRST DRIVE

SILENT STORMER

Subaru's 2.0 litre rally weapon takes liberties with the definition of high-performance. With more poise and less noise, it employs a tweaked turbo to post sub-15 secs for the 400m. MICHAEL STAHL checks the WRX specs

Forget tail-wagging wheelspin, axle tramp, induction roar, flailing fists, masses of metal being torn from rest. High performance used to mean offence to the eye, ear and nostril. Today it's a different world. Silent, compact, searingly efficient. Like a Beretta hidden in a purse. Or the Subaru Impreza WRX.

This apparently innocent sedan, built to carry Subaru's future hopes in world championship rallying, might just be the car to define high performance 1990s-style. It is of smallish to medium size (laughably tiny two decades ago), with quite a large engine of 2.0 litres (again, maybe one-third what was needed in '73). But with a techno-fest of multi-valves, turbocharging and constant four-wheel drive, it'll melt most of your 'modern' V8s quicker than a laser beam and even take the shine off a legend or two.

Very few cars can scythe from a standstill to 100 km/h in under seven seconds, or cover the standing 400 metres in less than 15. Australia's fastest ever muscle car, the Ford Falcon GTHO Phase III, was one of them. This little red wagon Impreza WRX – with just 1200 km under its still-tight belt – has already equalled that, with every sign that

the best is yet to come. Subaru's not the first or the only one to build high performance, turbocharged 4WD sedans and sports cars, of course, but among the Japanese manufacturers, its recent Liberty RS Turbo has rated above all but the Nissan Skyline GT-R in best getting the job done. The smaller, lighter Impreza WRX takes almost wholesale the engine and drivetrain formula of the Liberty RS, and improves upon it.

Which is all good news for us Aussies, because the Impreza WRX goes on sale here within the first quarter of 1994 at around $45,000 – equivalent to the price of the current RS. The WRX has to satisfy, rather than merely whet, our hi-po appetite before the stretched and re-styled Liberty range arrives a few months after it, because Australia won't be getting the storming, 184 kW/309 Nm RS version.

A ticket of $45K puts the Impreza WRX at a $12,500 premium over Mitsubishi's Lancer GSR, another all-paw turbo terrier that warrants comparison (especially since the 2.0 litre Galant VR-4 is no more). Both the WRX and GSR are similarly suspended, with MacPherson strut front ends and multi-link,

coil rears. Brakes, likewise, are ventilated front discs and solid rears, with ABS management.

Within their very similar dimensions, equipment levels are near identical. Their five-spoke alloy wheels even look the same, though the Subaru's are 15-inch diameter against the Lancer's 14s.

The WRX, with 1994 cm^3 against the Lancer's 1834 cm^3, even carries almost exactly the same kerb weight of 1200 kg. There the similarities – factual and philosophical – end. The WRX delivers more power and torque … differently. And it is also usefully faster … *differently.*

The WRX's turbo boxer four is not so different from that of its donor Liberty RS, with twin overhead cams per bank, four valves per pot and a single turbocharger with intercooling. Various tweaks have increased power to 155 kW at 6000 rpm (up from the Liberty's 147 kW at 6000), with a torque peak of 270 Nm at 4800 rpm (versus 260 Nm at 3600 rpm).

A lowered compression ratio and the torque peak occurring 1200 rpm higher both suggest that a bigger blower has been fitted to the WRX. The four-door sedan, with its aluminium bonnet, is some 155 kg lighter than the Liberty RS.

That helps the WRX's engine to get away with its shallow bottom-end. Mostly. In our tight test example, at least, the area south of the 3000 rpm border was declared a torque-free zone, with 3000 to 5000 rpm describing its happiest region for brisk suburban play. At this tempo, the too-wide and kinda funky gear ratios (virtually unchanged from the stronger-bottomed Liberty RS, but with first and fifth a shade taller) let it slip off the boost and dawdle like … well, like a normal car.

Driven thus, the WRX's spectacular surge and responsiveness are only really being hinted at. It begins to come alive approaching 5000 rpm, and between 5000 and around 6500 rpm (redline is 7000) the thing just *launches* if you even look at the throttle. With little accompanying engine noise – save the song of the turbines – the WRX slashes up the strip, each (spongy) upshift accompanied by a distant jolt through the drivetrain.

After a momentary mid-travel pause on the clutch to keep revs up, the WRX catapulted to 100 km/h in 6.9 seconds, 140 in 12.9 seconds (!) and flashed by the 400 metre mark at 150 km/h in a two-way

SUBARU IMPREZA WRX

Price:	$45,000 (est)
Engine	longitudinal flat 4
Induction/valves	turbo efi/dohc 16 valve
Capacity (litres)	1.994
Max. power: kW @ rpm	155 @ 6000
Max. torque: Nm @ rpm	270 @ 4800
Transmission	5 speed man
Drive	all-wheel
Tyres – size	205/55R15
Tyres – make	Michelin Pilot
Kerb mass (kg)	1200
Fuel tank (litres)	60
Acceleration (seconds)	
Standing start to 60 km/h	3.3
80 km/h	4.7
100 km/h	6.9
120 km/h	9.4
140 km/h	12.9
400 metres (km/h)	14.9 (150.1)
Rolling start	3rd 4th 5th
40-70 km/h	4.4 8.0 13.4
60-90 km/h	3.4 6.5 13.4
80-110 km/h	3.3 5.4 10.2
100-130 km/h	3.8 5.5 -

average of 14.9. The newness of the engine and a full tank prompt a punt on mid-14s when next we meet. For which we're counting the days …

In-gears times tell the tale of the WRX's big-boosted performance. Returning to our Lancer GSR comparison, the Subaru lags behind in every increment where its legs have yet to stretch. Only beyond 100 km/h in third gear does the Subaru outstrip the Mitsubishi's superior low-end flexibility. In fifth gear, cruising at 100 km/h, the tightly-geared Lancer GSR's turning at 3100 rpm – just above its torque peak. At the same road speed, the Impreza's tacho reads 2600 rpm – way short of its 4800 peak.

The WRX's brilliant chassis allows it – and *begs* – to be ridden like a two-stroke motocrosser. The penalty was a one-fill fuel consumption figure of 12.6 litres per 100 km – worse than we've had out of a Liberty RS.

Remarkably sharp, sensitive steering (with 2.7 turns) fairly whips it into all but the tightest of hairpins, where 4WD understeer eventually surfaces. Like its big brother Liberty RS, however, the WRX has a surprisingly agile and co-operative rear end.

With a lift-and-flick on entry, plenty of revs on board and a feathering of left-foot brake on exit to counter the mild, power-on understeer, the WRX slams through corners like a pinball off the bumper.

Such textbook 4WD handling and roadholding, helped by the highly impressive Michelin rubber, comes at no expense to ride quality. The soft suspension tackles harsh corrugations with a fair bit of suspension noise (and a shuddering bonnet), but absolute directional stability.

Braking power is immense and bulletproof, with strength and progressiveness in the pedal that refuses to go away. It's marred only on dirt by an over-sensitive ABS system that can't be switched off from the cockpit. That and the spongy gearshift action aside, the Momo wheel and hyperactive throttle when on-noise provide a landline to the road surface.

It all adds up to make the WRX a formidably grippy, Group N rally weapon straight out of the box. This red rocket just wants to get down, get dirty and get the hell out.

What you get into is an interior obviously not far removed from the regular Impreza. It's promised that equipment levels will be fairly high. That at least includes alloy wheels, front and rear driving/fog lights, fast glass, central locking, Nardi leather gearknob and the Momo wheel on a height-adjustable column.

Pedals are perfectly spaced for fox-trotting through the forest, but they're quite close to the seat (in long-armed, short-legged Italian fashion). Instrumentation is only adequate, even lacking boost or oil pressure gauges to fire the imagination. From outside and in, it's a Subaru Impreza to everyone else.

In fact, the Impreza WRX does without a lot of things in firing you into the forward view. Like wheelspin, sound and fury. Welcome to the *real* muscle cars of the 1990s – and your alarmingly immediate future. W

Subaru Impreza WRX

PICS BY ANDREW BRITTEN

Oh boy. And we thought the Liberty RS Turbo was fast. Picture the Liberty's brawn-infested boxer engine with more power and torque, placed in a car which is some 150kg lighter and even on paper it has to be quick. On tarmac, the Impreza WRX is simply roasting hot and what's more, in most circumstances its sheer composure and competence makes its performance entirely usable.

And it looks wild as well. None of the shrinking violet type of: "I-go-fast-but-you-wouldn't-know-it-until-you-try-me", understated Q-ship approach. Oh no. This Subaru has a big front air dam chock full of driving lights, black sill and bumper extensions, a dirty big wing on the bootlid and a letterbox-type bonnet scoop big enough to post naughty videos through. If that's not enough, try painting it screaming brilliant red.

The WRX exists to give Subaru a platform from which to launch its Group A World Rally Championship campaign, but if the road-going version we drove is anything to go by, then it should

Introverts need not apply for the newest 4WD turbo Subaru rally weapon. As Jonathan Hawley reports, the Impreza WRX bristles with machismo and has the performance to back it up.

SHORT 'N CURLIES	
PERFORMANCE	5
RIDE & HANDLING	5
NVH	3
EQUIPMENT	4
ECONOMY	4
STYLING	4
QUALITY	3
VALUE	4

Road Test

Road Test

win a swag of production-based Group N trophies as well. The concept is a familiar one: take a mild, somewhat boring, front-wheel drive sedan and plant in it the biggest, most powerful engine which will fit under the bonnet. Follow that by a computer controlled four-wheel-drive transmission to get all the power to the ground and unleash it on the buying public for homologation purposes. Then, if it starts winning rallies, use footage of it barrelling around Finland or wherever to sell more of the boring shopping trolleys. It's a concept done exceptionally well in the past by Audi, Renault, Peugeot, Lancia, Toyota, Mitsubishi, Mazda and plenty of other car makers.

Even if you couldn't give a flying fig about rallying, the WRX has an awful lot going for it. Not least is its pricing: at $39,990 it starts to look like a bargain considering that figure includes ABS, air conditioning, the completely overhauled drivetrain and other niceties such as a sunroof, power operated windows, aerial and locking. That price also puts it almost midway between the Mitsubishi Lancer GSR (which is less powerful and less well equipped) and Subaru's own Liberty RS. Gonna make for some painful buying decisions out there in Fantasy Land...

The body shell is basically Impreza sedan (although in Japan there's also a five-door WRX, would you believe) from which the normally aspirated 1.8-litre, SOHC flat-four is removed and dropped in the dump-master. Into the engine bay (accompanied by flashes of lightning, thunder bolts, spreading mist etc.) is crammed a hotter version of the same engine which powered the bigger Liberty RS Turbo. The boxer motor has just under two litres capacity, twin camshafts per bank of cylinders and each combustion chamber is ventilated by four valves. There's also electronic fuel injection and a single turbocharger which is fed chilled air through an air-to-air intercooler placed strategically under that bonnet bulge.

RIGHT: Styling is anything but subtle. ABOVE: Interior's grey tones, however, make it relatively subdued.

Power and torque outputs have both been raised over the Liberty RS's figures. It's now capable of 155kW at 6000rpm and 270Nm at 4800rpm (compared with 147kW at 6000rpm and 260Nm at 3600rpm) and the increases are due mainly to the IHI TDO5H turbo being about 10 per cent larger than the one it replaces, and its maximum boost has been lifted to 1.8 bar (11.8lb/in^2). The air/air intercooler replaces a water/air unit and it is less restricted, meaning turbo lag is reduced and allowed for the lift boost pressure. Meanwhile, in each cylinder head the cam lobes now act directly on bucket-type cam followers, meaning less friction losses within the valve gear.

These changes don't just improve the engine's absolute outputs, but according to Subaru help its driving characteristics through a wider spread of torque across the rev range, less turbo and therefore throttle lag, and even better fuel consumption. The torque curve bears this out in some respects, with 265Nm available at 3000rpm, 270Nm at 4800rpm and tailing off only slightly to 260Nm at 6000rpm.

Sounds good, but below 3000rpm the curve drops away sharply, and in practice there's little performance to be had until 2500rpm or so has been reached. That's partly a function of turbo lag, where slow moving exhaust gasses take more time to wake the sleeping turbo impeller, but we'd also bet there's a cam profile which favours high speed running. Because keep the foot buried and all hell breaks loose. The tacho needle blurs

Subaru Impreza WRX

LEFT: Two-litre flat four with turbocharger and associated plumbing doesn't leave much room under the bonnet. Note the intercooler towards the top of the picture which is fed directly by the large bonnet scoop.

across the instrument panel and WARRRUPPPSs into the revs limiter and it's time for another gear. There's no wheel spin, no axle tramp or even torque steer, just an absolute feeling of kilowatts being transformed into kilometres per hour.

So it helps to keep this engine on the boil, for that comparative lack of low-down urge is about the only downside to the WRX's operative ability. It's all-wheel-drive layout means traction simply is not a problem on dry surfaces, and can

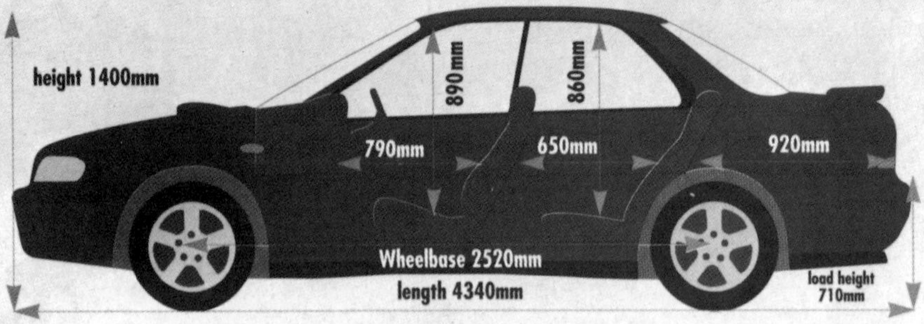

MODEL Subaru Impreza WRX

BODY............Four-door sedan
- Seating Capacity....................Five
- Warranty 3 years/100,000km
- Base Price $39,900
- Options fitted ....Nil
- Price as tested $39,900

ENGINE
- Location ....Front
- Cylinders ....Flat four
- Bore/stroke ....92.0 x 75.0mm
- Capacity ....$1994cm^3$
- Valve actuation....................DOHC
- Valves per-cylinder ....Four
- Induction ....EFI, turbo
- Compression ratio..................8.0:1
- Power ....155kW at 6000rpm
- Torque...........270Nm at 4800rpm
- Specific output 77.7kW/litre
- Weight/power ....8.0kg per kW

TRANSMISSION
- Gearbox...........Five-speed manual
- Driven wheels..........................All
- Ratios:
- 1st.......................................3.454
- 2nd.......................................1.947
- 3rd.......................................1.366
- 4th.......................................0.972
- 5th.......................................0.738
- Final.....................................3.900

BODY
- Kerb weight (kg).............1245kg
- O/A length 4340mm
- O/A width 1690mm
- O/A height 1400mm
- Wheelbase 2520mm
- Front/rear track ..1465/1455mm
- Fuel tank capacity ....60 litres

SUSPENSION
- Front...........MacPherson strut with lower wishbone and stabiliser bar.
- Rear....Independent struts with dual links and stabiliser bar.

STEERING
- Type.......................Power assisted rack & pinion
- Turning circle.......................10.4m
- Turns lock-to-lock...................2.7

BRAKES
- Front ....Ventilated discs (ABS)
- Rear...........................Discs (ABS)

WHEELS & TYRES
- Tyre make..........Michelin Pilot SX
- Tyre size 205/55 R15
- Wheel material ....Alloy
- Wheel size ....6J x 15 inch

PERFORMANCE
- 0-100km/h ....7.0 secs
- Standing 400m ....14.0 secs

FUEL CONSUMPTION
- CAR Australia average on test (l/100km).................10.9

often even overcome the engine's output in full-bore standing starts. This results in either a complete bog-down when the clutch is dropped, or said rotating disc starts to meltdown at 6000rpm if it's slipped a little. Our performance figures were, in fact, achieved on a damp surface which gave just a little wheel spin to help get the car cleanly off the line. How cleanly? Well, in just seven seconds it had passed 100km/h, and in 9.6 seconds achieved 120km/h. The 400 metre mark was crossed in just 15.0 seconds, and these times were notable not just for their speed but for the ease with which the WRX achieved them. There's a whizzing noise from the turbocharger, a sort of flat drone from the flat-four, a couple of gear changes, a shove in the back and its all over. The

277mm front/266mm rear disc brakes (which come standard with ABS) haul it back down to a standstill time and again without thinking of fade.

This sort of performance comfortably outguns even the hottest factory versions of Holden and Ford's V8s, so it would be expected that fuel consumption also rivals the grunters. Actually, it doesn't: over 950km from Canbera to Melbourne via some sensational high-country roads, the WRX averaged 10.9 litres/100km. The worst figure (which included performance testing) was 11.8 litres/100km; the best (which included a three-hour cruise up the Princes Highway) was 10.4 litres/100km. Impressive stuff.

We've always been a big rap for the Impreza's chassis with its soft, compliant ride and competent handling characteristics. They're taken to new highs in the WRX because unlike some cars, in which retro-fit 4WD is used to cure handling woes, in this case it adds another dimension, and on any type of road surface. On twisting, sealed roads there is simply so much grip from the low profile (although not excessively wide) Michelin Pilot tyres. It's possible to explore cornering speeds which start to reach the levels of the ridiculous, and the only ultimate vice is progressive and predictable 4WD understeer. And not a jot of the Impreza's excellent ride quality has been lost; absorption of even quite badly broken road surfaces is good and, if anything, the ride has been improved by firmer damping which helps reduce body roll and float. Noise levels are only reasonably low; the engine note is ever-present, road roar is intrusive on anything but the smoothest bitumen, but wind noise is at least relatively subdued.

It's loose road surfaces for which the WRX has been primarily designed, and the lower friction

Subaru Impreza WRX

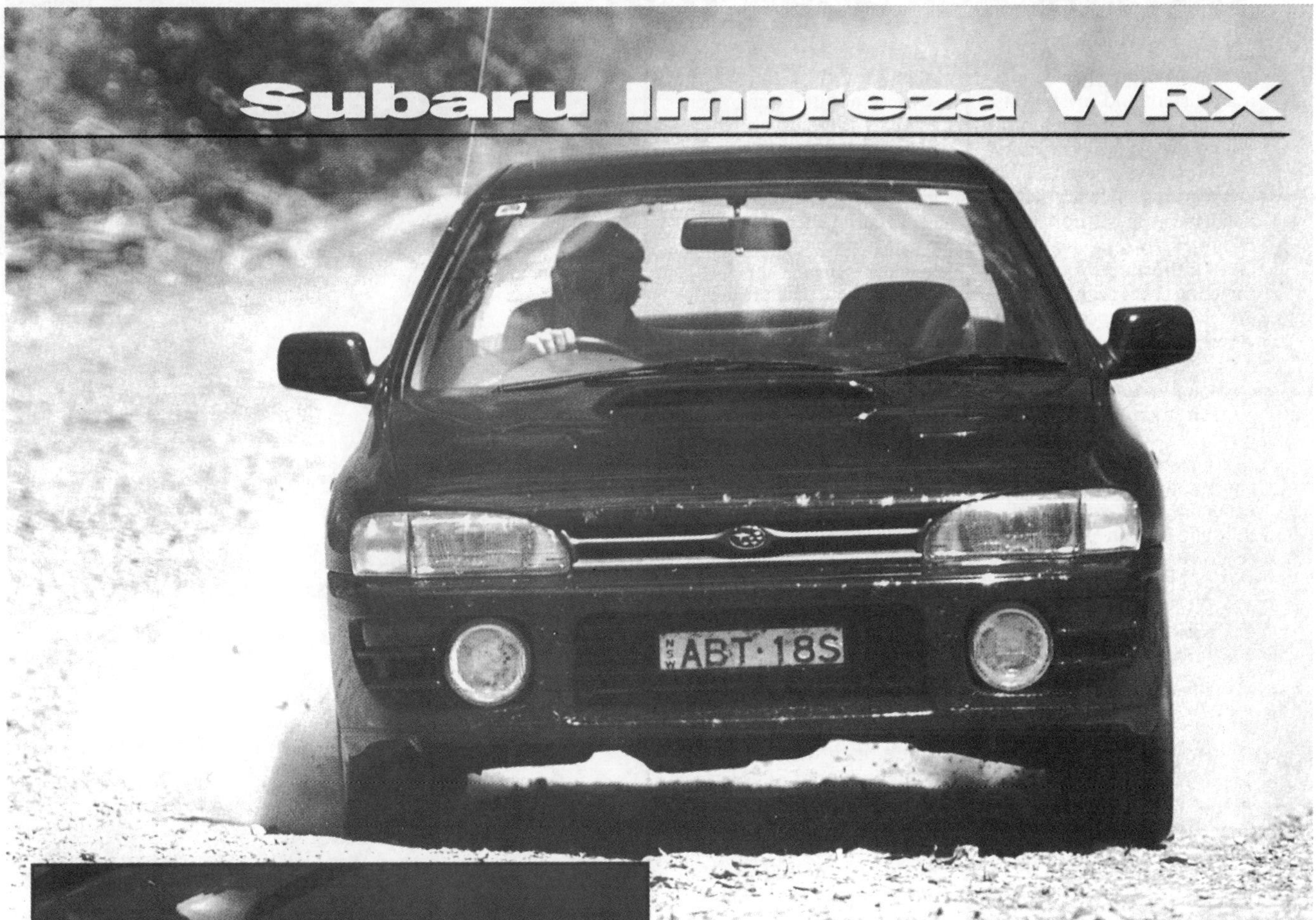

ABOVE: Loose surfaces are the WRX's favourite place; big power output and torque-sensing diffs make its attitude almost infinitely adjustable. LEFT: Spot the Impreza under all the gear. Note slotted vents for exit of super-heated air.

coefficient of our gravel roads show another side to its character. Simply, it's brilliant and very, very quick. First, the high levels of traction mean very little forward motion is lost in wheelspin, and it's educational to experience full-bore starts on loose gravel which have the car launching easily instead of digging holes. Second, it's a very forgiving chassis and powertrain combination because while it can act like a rear-drive car — slid sideways into a corner with a tail-out attitude — keep the throttle nailed (being very careful to have the right gear selected for proper on-boost characteristics) and the centre viscous-differential transfers torque to the front wheels which take over the power deliver duties to pull it smoothly out of the turn. It's predictable, great fun, and very safe.

In Australia the WRX is available as a sedan only, and a practical little car it is too. The boot is quite large, and the rear seats split 60/40 for even greater luggage capacity, but otherwise the back seat is fine for two passengers. The front seats don't look overtly sporty, trimmed as they are in a nondescript grey tweed, but they've been reshaped for more lateral support, and they're soft enough to be comfortable but supportive enough to cover the car's big cornering forces.

The spread of dashboard and instrument panel is mostly pure Impreza apart from a speedo calibrated to 260km/h, and a tachometer redlined at some 7000rpm. Interestingly, there's no form of boost gauge at all. The standard placky steering wheel has been replaced with a four-spoke Momo, while the Nardi parts bin has been raided for a leather-trimmed gear knob. All the switchgear follows foolproof Japanese practices, apart from it being difficult to determine whether the aircon button is on or off...perhaps a little light would help.

Summing up the Impreza WRX's abilities is pretty easy. It makes plenty of so-called sports cars look silly, not only because it is faster, but it handles, grips and rides with absolute aplomb. Plus it carries four people and luggage. Looks pretty good too. And it's cheap enough, considering all that performance. What else could you want?

CAR

The Autocar Road Test

IT ALL STARTS AS A staccato, burping hiccup of a roar, distant and mysterious through the forest gloom. Your pulse rate will rise quite involuntarily. Seconds later a screaming, spitting white metal monster will hurl itself, usually sideways and quite often airborne, through the shivering firs. This is rallying, one of motorsport's most visceral events, and for decades now one of the best ways for manufacturers to invest mundane motors with unspeakable glory. Lancia's Delta would be a forgotten name if not for the legendary Integrale, and who'd have thought a plain old Escort could be as painfully desirable as the RS Cosworth?

And while Subaru's Impreza offers up a pretty yawn-making road presence, the full-house rally version is about as dull as taxi ride through Beirut.

Now, for a non-princely £17,499, Subaru will deliver an Impreza Turbo 2000 4wd that strikes a rallying pose while threatening Integrale-busting performance.

A split second after turning the key you'll know this Impreza isn't bluffing; the free-revving engine possesses a bassy, coarse voice that says you'd better believe the 208bhp, 201lb ft power and torque figures in the brochure. However, those figures are produced at 6000 and 4800rpm respectively, demanding that the horizontally opposed four sings hard for its supper.

The Impreza's turbocharged and intercooled powerplant is effectively a mildly tweaked version of the four-cam 197bhp engine that serves in the Legacy Turbo, itself no slouch.

You won't find much power spilling uselessly on to the pavement, either, with a viscous-coupled four-wheel drive chassis that senses when a wheel is about to let go and sends the horses elsewhere. In normal, dry-road motoring, the torque split is about equal, front to rear. A relatively short-geared five-speed manual gearbox delivers the urge to grippy 205/55 tyres that ride on 15in rims. And the whole package weighs just 1213kg, giving the Impreza a better power-to-weight ratio than even the mighty Integrale.

Launch this unlikely junior supercar off the line in absolutely ideal conditions — a soaking wet track on a cool morning — and a scarcely believable 5.8sec later you will be travelling at 60mph. Even though the Impreza will do it all without leaving second gear, that is still a damned impressive result, one that tops the Escort Cosworth, Integrale and, just to shut the snide boys up, a Porsche 968 as well.

As we have already intimated, though, the 0-60mph figure is by no means a definitive performance yardstick, and while the Impreza will stay with the Escort Cosworth and Integrale up to about 75mph, from that point forward the other two will pull out in front. Given a little time to catch up, the keen-hearted Impreza will match the best with its 137mph top speed. And in the 30-70mph sprint the Impreza impresses with its 6.6sec figure, as good as the Integrale or Cosworth.

The Impreza behaves a little like its big brother, the Legacy, with turbo lag evident at low revs and less electric throttle response than its high-torque competitors. From about 3500 up to its 7000rpm red line the Impreza delivers a smoothly consistent rush of turbo thrust. It is an addictive sensation and one that will soon have you forgiving the engine for its somewhat coarse, unrefined manners.

You may find it less easy to forgive the gearbox, which is slow and rubbery at best and can baulk badly if you try to hurry it along.

Price as tested £17,719 **Top speed** 137mph
0-60mph 5.8sec **Mpg** 18.7

For Fabulous performance, good value for money, excellent warranty
Against Dull grey cabin, baulky gearchange, thirsty engine

The Impreza Turbo isn't just a straight-line thrill merchant. Responsibly doling out portions of its 208bhp to the most deserving wheels, it provides safe handling through fast, tight corners. That sense of security is helped out by the excellent grip offered up by the Michelin Pilots. Steam into a corner too quickly and the Impreza will start to run wide, with gentle and predictable understeer sending an early warning of relinquishing grip. Raising your right foot a shade tightens the line accurately and effortlessly, though. Simply put,

Subaru Impreza 2000 Turbo

There's no doubting the rally car in its blood, not least because it can leave a Cossie or Integrale standing

the Impreza feels confidently neutral, benign even, in most situations. With uprated springs and dampers, it corners flat too, but string a series of fast bends together and the Impreza can begin to feel unsettled, as if it needs a little more time to recover from the last corner. And the steering, while precise, is a touch too light for a car of the Impreza's meaty aspirations.

Ultimately, the Impreza is an easy car to drive quickly on challenging roads, but really keen drivers might resent the lack of involvement in the process; this Subaru is quite content to let its competent four-wheel drive chassis take all of the credit for a tight, well-executed corner.

The suspension's added stiffness does no favours for the ride, though. Over country roads with quick undulations and the usual blemishes the Impreza pogos about in a nervous and unsettling manner. The picture is a bit brighter on the motorway, where occasional disturbances are dealt with swiftly without any reverberations through the bodyshell.

Drive the Impreza the way its spoilered, scooped and vented bonnet suggests and the fuel pump penalty is severe. We managed a touring economy of just 26.7mpg and overall consumption ▶

The Autocar Road Test

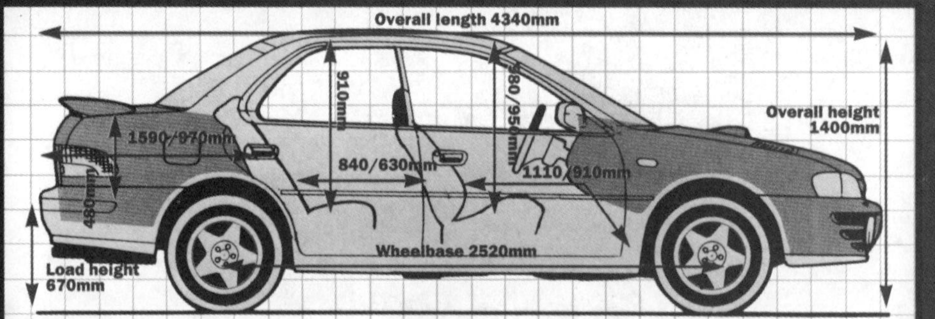

Width 1690mm
Front/rear tracks 1460/1450mm
Turning circle 10.4m
Interior width front/rear 1410/1410mm
Max/min boot width 1360/980mm
VDA boot volume seats up/down 0.35/0.60cu m
Cd 0.37

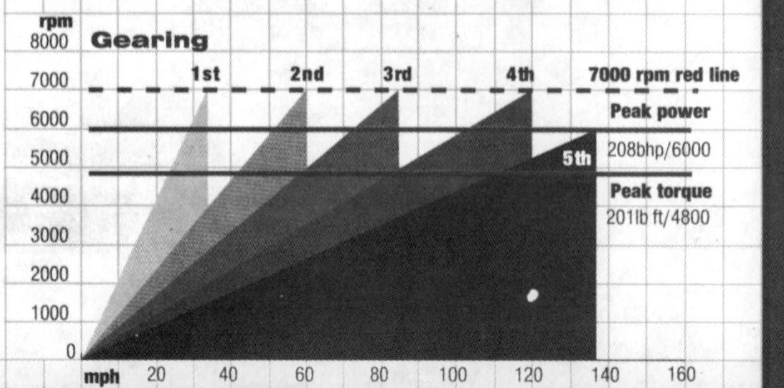

What it costs

Prices

List price	£17,499
Total as tested	£17,718

Equipment

(prices in **bold type** denote option fitted to test car)

Electric sunroof	£799
Driver's airbag	★
Metallic paint	**£219**
Adjustable seatbelt height	★
Seatbelt pre-tensioners	–
Variable speed intermittent wipe	★
Electrically adjustable mirrors	★
Auto-reverse radio/cassette player	★
Number of speakers	4
Remote hatch/boot release	★
Split/folding rear seat	★
Power steering	★
Alloy wheels	★
Adjustable steering column	★
Electric windows all round	★
Central locking	★
Height and tilt adj driver's seat	★
Driver's seat lumbar adjustment	–
Anti-lock brakes	★
Anti-theft system	★
Front foglights	★

★ standard – not available

Insurance group 17 (est)

Warranty

Three years/60,000 miles, six years corrosion, three years roadside recovery

Servicing

Major 15,000 miles, 4.25hr, parts £42.50 **Interim** 7500 miles, 2.2hr, parts £42.50

Parts costs

Set spark plugs	£60.00
Brake pads front	£88.12
Brake pads rear	£85.00
Exhaust (exc cat)	£802.52
Door mirror glass	£90.47
Tyre (each, typical)	£153.15
Windscreen	£203.96

Specification

Engine

Layout	4cyl, horizontally opposed, 1994cc
Max power	208bhp/6000rpm
Max torque	201lb ft/4800rpm
Specific output	104bhp per litre
Power to weight	171bhp per tonne
Installation	longitudinal, front, four-wheel drive
Made of	aluminium head/block
Bore/stroke	92/75mm
Comp ratio	8.0:1
Valves	4 per cyl, dohc
Ignition and fuel	electronic ignition, multi-point fuel injection

Gearbox

Type 5-speed manual
Ratios/mph per 1000rpm
1st 3.45/4.8 **2nd** 1.95/8.6
3rd 1.37/12.2 **4th** 0.97/17.2
5th 0.74/22.5 **Final drive** 4.11

Suspension

Front struts, coils, transverse link
Rear struts, coils, transverse link, trailing arms

Steering

Type power-assisted rack and pinion
Lock to lock 2.8 turns

Brakes

Front 228mm disc
Rear 230mm disc
Anti-lock yes

Wheels and tyres

Size 5x16in
Made of alloy
Tyres 205/55 R15 Michelin Pilot HX
Spare space saver

Performance

Maximum speeds

Top gear 137mph/6000rpm
4th 120mph/7000 **3rd** 85mph/7000
2nd 60mph/7000 **1st** 34mph/7000

Acceleration from rest

True mph	sec	speedo mph
30	1.7	33
40	3.0	43
50	4.4	54
60	5.8	65
70	8.3	77
80	11.1	88
90	13.7	98
100	18.7	108
110	24.8	120

Standing qtr mile 14.7sec/92mph
Standing km 27.5sec/114mph
30-70mph through gears 6.6sec

Acceleration in each gear

mph	top	4th	3rd	2nd
10-30	–	–	6.7	4.1
20-40	16.2	9.9	5.4	3.0
30-50	15.3	8.8	4.2	2.4
40-60	15.1	6.8	3.6	2.7
50-70	13.4	5.9	3.9	–
60-80	11.6	6.7	4.3	–
70-90	10.2	7.5	–	–
80-100	11.8	–	–	–
90-110	–	–	–	–

Fuel consumption

Overall mpg on test	18.7
Best/worst on test	26.7/9.5
Touring*	26.7mpg
Range	356 miles

Govt tests (mpg)
urban 24.8 **56mph** 39.8 **75mph** 30.1
Tank capacity 60 litres (13.2 gall)

Brakes

Distance travelled under max braking
Track surface wet
Anti-lock yes

30mph	11.3m
50mph	31.6m
70mph	62.8m
st qtr mile (92mph)	104.0m

Fade tests
Consecutive brake applications at 0.5g retardation from st qtr terminal speed
(figures on the right represent pedal pressures)

50lb
40lb
30lb
20lb
10lb
0lb

Weight

Kerb (incl half tank)	1213
Distribution f/r	54/46%
Max payload	550kg
Max towing weight	500kg

Sold by

Subaru UK Ltd, Ryder Street
West Bromwich, West Midlands B70 OEJ.
Tel: 021 522 2000

* Achieved over a pre-set test route designed to replicate an average range of driving conditions. **The figures** were taken at Millbrook proving ground, with the odometer reading 3500 miles. Autocar & Motor test results are protected by world copyright and may not be reproduced without the editor's written permission

1 Ventilation mode selector **2** Heater temperature control **3** Digital clock **4** Radio/cassette **5** Wipers stalk **6** Engine coolant temperature gauge **7** Revcounter **8** Speedometer **9** Fuel gauge **10** Electric mirrors switch **11** Headlights/indicators stalk **12** Ventilation inlet **13** Electric windows switches

Subaru Impreza 2000 Turbo

◀ of 18.7mpg, which is, frankly, poor when cars like the Cosworth will manage at least 20mpg overall.

The initial bright promise of the Impreza's astonishing performance is further let down once the light, pillarless doors are slammed shut, ensconcing occupants in an unrelentingly grey cocoon of nastily textured cloth and indifferent quality plastics. The height-adjustable driver's seat is supportive in most situations, and a fine driving position is helped out by an adjustable steering column, but this cabin's monochrome blandness is completely at odds with the car's shocking pink performance.

Also, those pillarless doors contribute to an irritating amount of wind noise at motorway speeds.

However unpleasantly executed the Impreza's cabin is, it does have plenty of Sainsbury's sensibility. The boot is an average size for the class, but folding rear seats almost double its load capacity. Front headroom is generous, especially when the pricey £799 sunroof option is left off the check list. Rear legroom is surprisingly good, too, but a low rear roof lining spoils headroom for even the modestly tall.

At £17,499 for the saloon and £17,999 for the oddly aspected estate, the Impreza Turbo offers more practical performance than anything else for the money (an Escort Cosworth costs £22,500 and a Delta Integrale £25,000). On that score alone, the Impreza is great value for money, but Subaru adds a generous specification to the mix; a driver's airbag, electric windows, alloy wheels and central locking are all standard. The only sour note, and it is a fairly rude one, is the knowledge that the Impreza Turbo is likely to cost a small fortune to insure.

Even though the Impreza fails to convince as a standalone champion in the mould of a Cosworth or Integrale, Subaru's little rocket is still a stunning accomplishment. An enthusiast with less than £18,000 could never hope to find himself behind the wheel of a sophisticated four-wheel drive chassis backed by Cosworth levels of performance. Not until the Impreza Turbo 2000, that is. ■

Impreza Turbo doesn't skimp on the goodies; central locking, electric windows and driver's airbag are all standard fit. Cabin's unrelenting grey is a bit of a downer, though, and the upholstery cloth has an industrial texture

Cabin space is good, although rear headroom is a little tight for the modestly tall. Rear seats fold to nearly double the load space

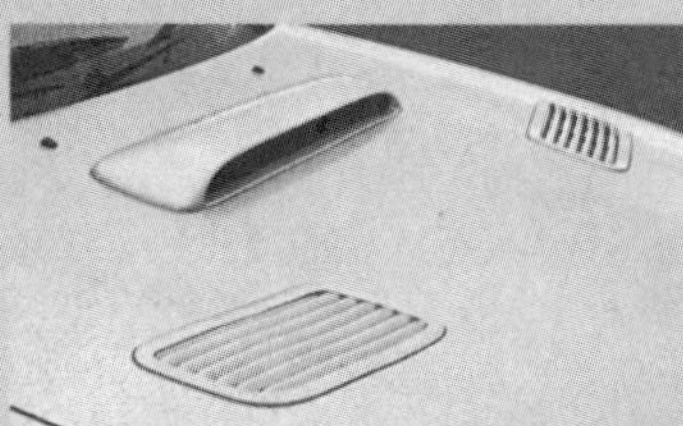

Impreza announces serious intent with dinner-plate-sized foglights and more scoops and vents than an F16. Looks good, but insurers aren't likely to be so keen

Engine is a mildly tweaked version of Legacy Turbo powerplant. It delivers stunning performance but isn't the most refined. Fuel consumption is on the high side, too, especially when pressing on

Massively secure handling aided by four-wheel drive and grippy Michelins. Very quick point to point

The Chosen Phew

It's a limited edition, so both you and it are going to have to be quick. Subaru's four-wheel-drive, rally-bred Impreza takes one step closer to the heart, courtesy of Prodrive's team of surgeons. Better than Cosworth and Integrale on ordinary roads? Greg Fountain drives one of the few

Photography by Peter Robain
(Special thanks to Castle Cement Ltd)

It is not an Escort Cosworth, and you would not mistake it for one merely by looking at it. And what's more, it is not an Integrale. But what Prodrive has done to Subaru's bred-for-the-rally Impreza turbo may be closer to the rally ideal than you'd imagine, so stay tuned. Don't flick channels merely because you can't envisage a Subaru key fob on the pub table in front of you — show some imagination.

One glance at Subaru and you'd be forgiven for thinking you were looking into a four-way mirror, because four distinctly different images come back. Is this the marque that builds first-class farm vehicles renowned the world over, or is it the maker of average saloon cars that we can't quite remember anything about? Or wait — is it the outfit that gave us the wild, wonderful but unloved SVX coupe? Or the name on the cars that have dominated the British Rally Championship for the last three years?

No wonder you can't envisage that key fob. It would take two hours to explain to your uneducated pals that you've invested in something worthwhile. Far better to take them for a spin in your Prodrive Impreza, after which they're unlikely to be in the mood for scoffing.

Banbury-based Prodrive is the team that prepares the British rallying Subarus, its success having begun with the triumphant Legacy of Colin McRae, and gone on to encompass the 22-year-old current champion Richard Burns. Its work on the limited edition Impreza you see here amounts to a good deal more than the absolutely nothing at

The Chosen Phew

all which Williams contributed to the Clio. It amounts to suspension modifications, new wheels and tyres and extensive cosmetic interior mods. It amounts, in fact, to £3950 fully-fitted, giving the 25 special cars a price tag of more-or-less exactly £22,000. Just for the record, that's £2,240 less than the Cossie and £3000 less than the Integrale. My mission here is to find out whether less really does mean more, or whether my primary school teacher Mrs Vaughan was right to maintain that less does, in fact, mean less.

'It's more Integrale than Escort in corners, its body roll noticeable, its balance balletic'

Prodrive had a head start with the Impreza Turbo 2000 4WD, because it's a good car, and I rather suspect they had to do a whole lot less to it than they expected. It's already powered by a version of the 2-litre quad-cam turbo unit which is a legacy from the Legacy, and it's already revised to give more power (208bhp @6000rpm) and more torque (201lb ft @4800rpm). It pumps this Impreza's blood with undisguised verve, leaving nobody in any doubt at all that we're dealing with first-degree fastness. The journey to 62mph takes 6.6 seconds, and believe me, time flies.

How good is the engine? Well, Prodrive thought it was good enough already and left it alone. Almost totally devoid of turbo lag, partly courtesy of an air-cooled intercooler (as opposed to the old water-cooled item), it's a creamy, rorty deliverer of urge, its character more fruity than you'd expect of something heavily governed by an engine management computer worthy of NASA.

On the road, and coupled to the kind of four-wheel-drive set-up that woos farmers so successfully, it applies its power without a moment of hesitation.

The gearbox, however, is more disappointing than getting no Christmas presents. Despite the chance to fondle a charming carbon-fibre effect gearknob, you find your blistering progress too often obscured by wholesale wrestling bouts as the stick shies away from third or fourth. I tried the gentle touch and I tried impersonating the semi-automatic item's hyperactivity, but wound up frustrated. I was warned about it beforehand, and I think Subaru is aware that it's a weakness, but it flaws the fun rather than obliterating it entirely.

● *Uprated Legacy engine gets air-cooled intercooler and 208bhp, and turbo lag has been knocked firmly on the head*

So, I hear you ask, if they didn't lift the bonnet, what exactly did the Banbury boys do? Well they looked under the wheel arches and sucked air through their teeth when considering the MacPherson-struts-all-round suspension set-up. Springs and dampers had already been uprated for the Turbo model, so they opted to uprate the set-up further, lower the whole caboodle and add an anti-roll bar. In addition, we get 7x16in alloy wheels shod with Pirelli P Zero 205/50 ZR 16 rubberware.

The result is a rock-hard ride, which is no bad thing for a car this nifty, but I fear they may have overdone it just a smidgeon. Certainly the chassis baulked somewhat at the prospect of country undulations, flicking the powered steering wheel from side to side in my hand and giving the impression that a wheel may have lifted once or twice through lack of pliability. This is the only area in which the Impreza feels less than totally secure, and perhaps the reason why it's not quite an Integrale.

But in the corners it sticks to its task as if there'd been an Evo Stick spillage, allowing thunderous entry speeds and fluent exits of the type that bring the Integrale back into your mind just when you were starting to doubt the Subaru's credentials. The Impreza is more Integrale than Escort when you're cornering, its body roll noticeable but its balance positively balletic. It urges you to trust it and you do, despite what your mother said about strangers.

You can trust the brakes too, the all-round disc system mated to four-channel ABS offering a measured, polite degree of stopping power guaranteed not to subject your passengers to seatbelt injuries even when the spring lambs are in leap-out-in-front-of-you mode.

Cosmetically, the Impreza has you on its side from the very start, its hungry looks and purposefully

● *Cornering is Impreza's number one strength — four-wheel drive gives sure-footedness, making your exit fast and fluent*

bulgey aluminium bonnet defying the lack of familiarity and questioning how the standard model can look so very, very plain. Inside it's even better, Prodrive having gone to town on the dashboard with bits of carbon-fibre effect plastic turning up all over the place. You are allowed to have a sports steering wheel too, but our car was fitted with the standard airbag-equipped item in a bid to make sensible concessions to safety. Laudable, yes, but the wheel could still be more compact, better looking and offer at least a shred of tactile fun. It sits most uncomfortably with that glorious gearknob.

The seats, fitted by Prodrive, are cloth-trimmed Recaros in risky colours that match the door trims, but they're so comfortable that you can forgive any visual transgression. They give the cabin a tight, neat, functional air — the look of a place where business and pleasure can get done at the same time. A word of warning about the immobiliser, however. It will certainly stop that idiot thief getting away with your motor, but it might also stop you getting to work in the morning, such is its determination not to let go. It reminded me of an over-zealous security guard who's known you for 25 years but won't let you in because you're not wearing your identity badge. Be warned. A getaway car it ain't.

A family car, on the other hand, it certainly is. With four doors, four proper seats and a big boot that gets even bigger when you fold the seats down, this is as good a way as any of transferring granny, Mary, Joey, Victoria, Bonzo, two rubber plants and a small occasional table to the garden centre so quickly that you might still get back in time for the *EastEnders* omnibus. There is some real compromise here between duty and the beast.

Prodrive has not set out to change the world by tweaking this car, but it has succeeded nonetheless in going the extra mile that Subaru opted not to go. Good though the Impreza Turbo 2000 4WD is, it feels a whole lot better after a trip to Banbury. Under the operating theatre lights it has undergone cosmetic surgery with pleasing results, and the best vote of confidence you could give to its heart is that the Prodrive surgeons didn't feel the need to operate further.

Not a Cosworth, because it doesn't have the all-round sympathy with the road that the winged Ford offers as standard, and not an Integrale, because it has neither the chassis fluency nor the presence of everyone's favourite Lancia, it is still a damn fine road car, a close approximation of rallying pedigree, reasonably priced and practical besides. Plus, Mr Jones doesn't have one.

If you feel the urge to buy a wolf, but would prefer it to be dressed in sheep's clothing, then look not even a little bit further. ❍

● Many thanks to The Deer Leap Garage, Ringshall, Berkhamsted, Herts. Tel: 0442 842273

● *Prodrive has done the business here, with excellent Recaros, loads of carbon-effect plastic and that gear knob*

● *Almost as much fun to be had in this Subaru as in an Integrale, and that's saying something*

L606 YWD

Pure genius

But not many guineas. Colin Goodwin takes Subaru's Impreza Turbo to Ireland, and finds it a tastier brew than Guinness

PHOTOGRAPHS BY COLIN CURWOOD

YOU DON'T NEED THE YELLOW-stars-on-blue signs to see where Eurodosh is being spent in Ireland: main roads are being resurfaced and widened into super-wide single carriageways, and towns now have wonderful new bypasses where previously only time passed by. Heavens, there are even a couple of motorways near Dublin.

We're not looking for smooth-surfaced super-speedways today, though. No, we're heading for the west coast – where bog roads ladder a landscape that for millions of years has borne the brunt of the Atlantic Ocean's wrath – and I'm expecting the new Subaru Impreza Turbo 2000 to provide huge entertainment.

Subaru's latest has already imprezzed me (there, that's the predictable pun out of the way – no more, I promise) on the run from London to Holyhead. As I frequently do, I've allowed myself an extremely Ron Hopeful amount of time for the journey and, to make matters worse, I stopped off at Knutsford services on the M6 to continue work on my first heart attack.

My previous Impreza experience was last summer – in a five-door like this one, but powered by the 1.8-litre flat-four. And flat it is. With a weedy 101bhp to power all four wheels, the car is a sluggard – dynamically good, but drastically short of puff. We knew back then that the turbo engine was on its way, and our experience of it in the Legacy gave notice that the Impreza Turbo was going to be a flier. What's more, the Impreza has a still more potent version of that motor – 208bhp instead of the Legacy's 197bhp. Yes, that's right, 208bhp: a huge 107bhp more powerful than the next Impreza down. That should cure the sluggishness problem.

Even this bolt of brawn isn't enough to solve my time trouble, however, for the last part of the run to Holyhead involves a tedious 19-mile slog across Anglesey – along roads where overtaking is almost impossible. Thankfully, I make the Sea Cat terminal just in time. My arrival is so dramatic that when I tell the enquiring security policeman my profession and mission, he remarks, 'I thought so, by the way you made your entrance.' The Sea Cat to Dun Laoghaire takes just 1hr 35min, which is two hours quicker than the boat. If the sea is very rough it doesn't go, and if it's mildly rough you'll feel very rough.

The next day I meet photographer Colin Curwood at a wet and foggy Dublin airport. Curwood is most scathing about the Scoobyroo, laughing at its looks and doubtful of its prowess. It is an odd-looker, the Impreza five-door, but that's one of the things I like about it. The saloon looks dull, and could have rolled off any CAD/CAM, but the hatch, with its curvaceous rear door and shapely third side-window is unmistakably different. What has really offended Curwood's aesthetic eye is the collection of wings, skirts and bonnet addenda that adorn the puffed-up Impreza. Usually I'd agree with him, but this is a car whose big brother goes rallying, so the speed furniture doesn't seem ridiculously out of place.

Your view forward from the Impreza

Turbo's driving seat is part-obscured by the air-intake that sits atop the aluminium bonnet. I love air-scoops, but only if they're actually functional. This one is split 75/25, the larger part feeding the air-to-air intercooler (the Legacy Turbo uses a water-to-air unit), the smaller intake funnelling a cool rush of air to the turbocharger. Further down the bonnet are two Cossie Escort-like air-intakes. On the rally car these are highly functional, but those on the road car are for show only. Nonetheless, they have to be there for homologation reasons. Wrapped around the Sube's nose is a deep air-dam which contains two of the most enormous fog lamps you'll ever see. They look like those huge dentists' lights that one stared panic-stricken into as a child. Between them is a large radiator-cooling mouth, to whose gauze cover an ugly and restrictive numberplate is fixed. Leading backwards from the front air-dam, small, black, and sensibly bendy plastic skirts continue along the length of the car's sides – breaking only at wheel-arches that contain attractive five-spoke 15in alloys – to the rear bumper/air-dam. Two wings decorate the rear hatch – one at roof height, the other at waist level.

We leave Dublin on the N4, following signs for Galway. The roads are a mixture of normal-width country A-road, and the four-car-wide carriageways mentioned earlier. Once out of Dublin's sprawling suburbs, I get a chance to wipe some of the sneer off Curwood's face. To get the Impreza's large herd of horses to gallop *en masse,* you need to dial up at least 3000rpm, better still 3500rpm. But far from being annoying, the delay adds to the thrill. The flat-four sounds serious from low revs, but when you hear the turbo begin to whine, the engine note changes to a bark and the car leaps forward. Overtaking needs a bit of planning: leave a larger than normal gap between yourself and the car in front, then start the manoeuvre earlier than you usually would and use the run-up to build up the boost; then *rrrruffff* – you've gone.

The Gardai (the Irish fuzz) are thin on the ground in the countryside; besides, from past experience I've found that they are much more open to debate than our lot, much keener to leave it at a dressing-down. So progress towards Galway is swift,

Clockwise from top left: Impreza's interior looks cheap, but then car *is* cheap. Matt Molloy's is a traditional Irish pub, but author Goodwin hates Guinness. Irish weather doesn't surprise, but Impreza copes. Sunset in Connemara. Subaru has fantastic ride, laughs at bumpy bog roads

although I resist the temptation to verify the car's 142mph maximum speed.

Fellow drivers make no fuss about me and the bright-red Subaru galloping through their country, and move over to let us fly on our way. Throughout Eire (I've never been to the north, so I don't know if it's the same) whatever you're doing, whoever you are, there's a sense of freedom, a feeling of 'if it doesn't tread on my toes or harm me, then it's OK' that is fast disappearing in England. We – especially we motoring enthusiasts – are losing our freedom, and no-one seems to give a toss. It's for this sense of live and let live, and of course the beautiful countryside, that I love Ireland. It's certainly not for the climate, or for Guinness (which to me tastes and looks like creosote, but without the nice smell).

Surburban Galway is horrible, a mass of warehouse stores and ugly tower blocks. We're soon out of it, though, and heading out on the R336 along the Galway peninsular. The weather is atrocious, heavy rain whipped up to jet-wash intensity by Atlantic gales. Curwood is looking depressed, but is soon hanging onto the passenger's-side grab-rail with all thoughts of exposure, both photographic and medical, temporarily scared out of his mind. The sections of N-roads that have yet to feel the effect of an EC grant cheque are a bit rough, but the minor roads that run up the west coast are something else. The harsh weather breaks and cracks them, and the waterlogged bog underneath causes them to droop and sink to their own level.

The Impreza has the most amazing damping. Yumps, pot-holes, cracks and ruts fail to disturb it, evidence no doubt that Subaru's rallying experience is paying off. Before this trip I didn't think of this car as a serious rival to the Lancia Delta Integrale or the Ford Escort Cosworth, but now I'm changing my mind. Driving an Integrowler this fast across this moonscape would be terrifying, if not impossible. The wonderful Italian is so tautly sprung that much of the time would be spent airborne. Sure, I'll bet an expert rally driver could blast the Delta at enormous speed across here, but I'd wager that a normal person would be faster in the Impreza.

What's more, the Subaru is much more solid than the Lancia. The Japanese car's interior is certainly cheap-looking, and very plasticky, but it's free of squeaks and rattles. Punish a Lancia as we are punishing this Subaru, and it would issue a greater range of noises than a Yamaha synthesiser. The Subaru's Woolies interior doesn't bother me, though. One of the reasons why Subaru is a favourite maker of mine is that it has ignored fashion and steadfastly produced its own type of vehicle. Subarus are working machines, popular with farmers, and the last thing agricultural folk need is a vehicle whose cabin will decompose at the sight of a muddy welly.

The Subaru Legacy turbo is fitted with a

Clockwise from top left: fishing village of Roundstone lit by rare sunshine. Five-door Impreza looks wackier than saloon. Big hood-scoop feeds intercooler, but others are fake. 2.0-litre flat-four turbo gives 208bhp. Croagh Patrick, Ireland's holy mountain

wonderful Momo steering wheel and leather gearknob, but the Impreza has an airbag-equipped wheel instead of the Momo. Still, I'd rather eat bag than posh steering wheel. Nardi supplies the gearknob, so you still get a touch of luxury. As well as displacing the Momo, the airbagged wheel gives the steering a slightly heavy feel. I think it may well be partly psychological, but the Impreza's steering doesn't have quite the feel of the Legacy's system. That said, it's still a whole lot better than most Japanese cars' steering apparatus.

The rain continues to lash at us as we head north towards Connemara; there's standing water everywhere, occasionally so deep that the speeding Sube throws gallons up and over the windscreen. I love driving in this goo. As a motorcyclist, often in the past not by choice, I really appreciate the cosiness of a well heated car, and its radio and comfy seats. If you've not ridden bikes in appalling conditions, you'll find it hard to understand my sense of wonderment at travelling in comfort through Hades.

I wonder how John Alcock and Arthur Whitten-Brown felt as their Vickers Vimy touched down on a piece of flat ground near Ballyconeely on 15 January 1919, having left Newfoundland 16 hours 27 minutes earlier. I presume the weather was kinder that day, or the first crossing of the Atlantic by plane would have ended in disaster. But it didn't, and there's a large stone monument commemorating the two pioneers' achievement.

Subaru gives you a choice of drivetrains, from 1.6-litre 2wd saloons and five-doors, dual-range selectable 4wds with 1.8-litre motors, to the permanent-4wd turbo cars. The four-wheel-drive system uses a viscous coupling between fore and aft axles that apportions equal torque in dry conditions, but transmits power to the end with most grip in slippery conditions. There's also a torque-sensing diff on the rear axle that performs the same task for the rear wheels. The system works very well, too – it's practically impossible to provoke wheelspin, unless you're particularly mechanically unsympathetic.

At last it has stopped raining, and the sun is battling through the clouds, promising to give us a spectacular sunset. It's a brief respite, however, for as we roll into Westport for the night, the wind has gathered again, and the clouds have begun to hose the Emerald Isle yet again. And it's still raining hard as we retrace our steps back down the coast to Lecanvy where Ireland's holy mountain, Croagh Patrick, looks down on us through the mist as we photograph the Impreza on a quayside. The Sub-aru looks as though it might become Sub-marine, as wind-whipped seas crash against the land's defences, bursting over the quay's stone wall.

From there we head back through Westport and on up through Co Mayo towards Ballina. The sun is playing peek-a-boo, lighting up the vast expanse of peat bog as we blast across tagliatelle roads. If I screw up along here, we'll be pitched into the bog and will disappear without trace, to be dug up in thousands of years' time and given names like Eric and Fred, the Bog Men, and be put on display along with what's left of our vehicle, in some museum of natural history. We should be all right, though, because the Impreza s chassis is so forgiving, so surefooted, that it can make allowances for any ineptness on my part. If you're really ham-fisted and throw the Impreza into a sharp corner with unreasonable brutality, then it will understeer straight on until you lift off and allow it to tuck back in. Drive sensibly, and the levels of grip will amaze you. Our car is running on 205/55 Michelin Pilots, which have excellent roadholding in the rain. The Scooby's brakes are excellent, too, making slowing down almost as exciting as speeding up. The vented discs both front and rear are backed up by ABS, which I manage to activate only when landing from a yump before a corner.

Curwood has a plane to catch today, and I must head back to Dun Laoghaire for more stomach-churning, so we must turn inland onto the better roads and head for Dublin. It's a great shame because what I really want to do is turn around and go back down the west coast, right the way down to Cork. There's so much to see and the Subaru Impreza Turbo has proved to be such a good conveyance from which to see it.

There are a few areas where the Sube is let down, one being a gearchange that is clumsy and tricky to work quickly – which is a traditional Subaru failing. The other problem is its rather WC Fields-like thirst, which combined with our somewhat speedy progress resulted in an 18.7mpg average. Not very good, although I'll bet it improves dramatically when you drive more normally.

On smoother roads you'd probably have more fun in either a Lancia Delta Integrale or a Ford Escort Cosworth. However, I haven't mentioned the important bit yet – and that is what this thing costs. A figure of £17,999 sounds reasonable to me. In fact, in my opinion you'd be extremely hard prezzed (oops, sorry) to find a more entertaining or stronger-performing car for less money.

Rare bird

The Japanese have long been masters of state-of-the-art cars. But they haven't always been able to blend technology with character. There are, however, some honourable exceptions . . .

Despite its current prominence in international rallying (and let's face it, Colin McRae, in particular, has sometimes aroused the habitually disinterested national press), Subaru's road cars never have had a particularly high profile in the UK.

The last example of the breed we tested (in February 1993) was the 2.0, turbocharged, four-wheel drive Legacy, since usurped on Forestry Commission land by the more compact (and already successful) Impreza Turbo 2000 4WD, though the Legacy range continues to be available for road users.

Simply *looking* at the road version of the Impreza is enough to make you think that it is already Subaru's most desirable model. Available as a roomy four-door saloon (tested here) and five-door hatch/estate, it is prettier than the Legacy, but it retains the latter's Q-car charm. Only the aggressive front end treatment (bonnet louvres and giant fog lamps) offer any clues to the car's true purpose.

But. . .

With permanent four-wheel drive and a top speed of 140 mph, its obvious rivals, on road and stage, are the Ford Escort Cosworth, the departed, but still magnificent, Lancia integrale and the Toyota Celica GT4. The last three are all similarly potent, yet all are far more expensive.

Why?

To some extent, you get what you pay for. Although the Impreza's cabin isn't exactly basic, neither is it remotely inviting. Equipment levels are adequate, save for the absence of Recaro-standard seats, which is most noticeable when the Gs are loaded up. While the standard items are comfortable – and a good driving position is attainable, even for tall drivers, via the height-adjustable steering column – the absence of side bolsters and inadequate support under the knees is all too apparent. Accommodation is generous in all areas and five adults can comfortably be located. You can't say that for some of the Impreza's rivals.

The workmanlike interior, dominated by the ugly, air-bag retentive steering wheel is typical of old-school Japanese design, with rather dull tones of grey cloth offset against cheap plastic. Although the controls are ergonomically sound and the instruments perfectly legible, there are even fewer internal clues to the Impreza's pace than there are external. As an aside, don't leave valuables in this car – there is no security alarm, though an immobiliser is standard.

If you want hard evidence of performance, look beneath the lightweight bonnet: you'll find a longitudinally-mounted, aluminium, flat four, quad-cam, 16-valve engine which has been tweaked considerably since it first saw service in the Legacy. The upgrade includes an air-cooled intercooler, which has whipped power up from 197 bhp to a more competitive 208/6300 rpm. Likewise, torque is up from 193 lb ft to 201/4800 rpm. With these mods and the Impreza's lighter, more rigid bodyshell, Subaru claims 0-62 mph acceleration in 6.6s (significantly quicker than the Legacy), but our test car actually *felt* even faster. Sadly, time constraints prevented our being able to conduct a fully calibrated check.

The five-speed gearbox is similar to the Legacy's, but the four-wheel-drive system has been much revised. Torque is continually varied, laterally and longitudinally, according to available grip in the prevailing conditions.

The chassis is barely different to that on lesser Imprezas, featuring little more than uprated dampers and springs. MacPherson struts all round, combined with trailing arms at the rear, form the basis of the suspension set-up. Needless to say, for a car with this performance, the front brake discs are ventilated and assisted by a four-channel ABS system. Engine-speed related power-assisted steering is standard.

If the Legacy was not as inspiring as a Sapphire Cosworth to drive, the Impreza's smaller dimensions certainly promise more.

However, when you examine the prerequisites of 'driveability' (responsiveness, a flat torque curve and wide power band), the Impreza's flat four doesn't exactly appear tailor-made, leastways on paper.

It is rare for even the most advanced turbocharged engine to provide razor-sharp responses or pull from low revs, but at least

ENGINE	
Location	longitudinally front-mounted
Cylinders	four, horizontally opposed
Bore × stroke	92 × 75 mm
Capacity	1994 cc
Compression ratio	8:1
Valve gear	dohc, four valves per cylinder
Power	208 bhp/6000 rpm
Torque	201 lb ft/4800 rpm
Fuel	unleaded, 95 RON

TRANSMISSION	
Type	five-speed manual, permanent 4wd

GEARBOX		
Gear	ratio	mph/ 1000 rpm
First	3.454:1	n/a
Second	1.947:1	n/a
Third	1.366:1	n/a
Fourth	0.972:1	n/a
Fifth	0.738:1	n/a
Final drive	3.545:1	

SUSPENSION	
Front	MacPherson struts, coil springs, L-shaped transverse link
Rear	MacPherson struts, coil springs, transverse dual links, trailing arms
Wheels	alloy, 6JJ × 15
Tyres	205/55 R15

BRAKES	
Front/Rear	ventilated discs all round, ABS

STEERING	
Type	rack and pinion, power assisted
Turning circle	10.4 metres

DIMENSIONS	
Wheelbase	2520 mm
Front/Rear track	1465/1450 mm
Overall length	4340 mm
Overall width	1690 mm
Overall height	1400 mm
Kerb weight	1235 kg
Fuel tank	13.2 gallons

PERFORMANCE	
0-62 mph	6.6s
Maximum speed	142 mph

Figures supplied by Subaru UK

FUEL CONSUMPTION	
Average for test	28.3 mpg
Government figures:	
Urban	24.8
56 mph	39.8
75 mph	30.1
LIST PRICE	£17,499

Cabin is typical of modern Japanese thinking, but the overall package is endowed with rare character. Unusual flat four engine revs freely.

the Impreza's delivery is pretty smooth and it becomes ever more willing as the revs rise.

That said, there are noticeable phases in the unit's progress. Floor the accelerator and initial lag is superseded by a sudden kick when the turbo whirrs in at around 2500 rpm. Then, at about 5000 rpm, when you'd expect the power to be tailing off, it gets its second wind, pulling strongly to the red line. If you're caught off-guard in a high gear at low revs, then no amount of throttle stamping will bring anything other than a tardy response. This is an engine which thrives on being revved and is unwilling to idle at the behest of a lazy driver. However, if you wish to make less frantic progress, the four-pot is flexible enough to hold its own.

The chassis dynamics *will* bring a smile to your face. The Impreza is as easy to drive as it is on the eye, a sharpish clutch being the only noticeable concession to power. The brakes can grab a little if used without sensitivity at low speeds, but otherwise the short nose and ample vision, assisted steering, excellent driving position and amazingly light gearchange make urban driving a doddle. Refinement cannot be faulted. There's little wind noise, but enough sound from the engine to keep you awake on motorways, and everything bar obvious pot holes are adequately absorbed. Surely something that rides this well cannot be sporting enough to be entertaining?

Wrong!

Show the Impreza some interesting roads and the sweet flat four's novel bark implores you to remain closer to the rev limit (7000 rpm) for much longer than would be to tolerable in, for example, a Cosworth (redlined at 6000). This is no roller skate – as the body roll consistently reminds you – but if you take the Impreza by the horns it will obey your commands in rewarding fashion. Forget about sloppy understeer (though it does so, mildly); only well-sorted tiddlers such as Renault's Clio Williams can offer significantly better turn-in.

The short wheelbase makes the chassis very throttle-sensitive, and one feels that the car would perform better on gravel, which requires a more flowing driving style than tarmac. When confidence is up and vision is ideal, great fun can be had setting the car up for a series of tight corners – even with its high level of grip.

If you're quick with the light steering, you can have oodles of fun without undue drama. Subaru has retained (just) enough steering communication and kept extremes of lock to a compact minimum. The nose will tuck-in if you lift-off when cornering at speed, and the tail returns to the chosen line with a little more throttle application and, possibly, a whiff of opposite lock if you've gone way over the top.

Its poise and grip isn't limited to the tight stuff. The Impreza's mile-eating capability over super-fast terrain is almost awesome. There aren't many cars the sane side of 50 grand that could leave an Impreza far behind; most of those that could would be hard-pushed to keep up on tighter, narrower roads.

Stopping this formidable machine also inspires confidence, though it is no match for an Escort Cosworth in the braking department.

The transmission is the main dynamic flaw: the gearchange is not the smoothest we have encountered, and there is often a trace of clutch judder. Also, there was an intermittent tendency for the engine to behave as though it was being starved of fuel if ever you backed off the throttle abruptly.

Verdict

The Subaru Impreza Turbo 2000 4WD is a no-nonsense performance saloon, utterly devoid of frills and gimmicks. Yes, the interior is dull, but at least it is comfortable. If you're looking for something a little more extrovert, the aforementioned rivals will further demolish your bank balance to the tune of anything between £7000 and, in the case of the Celica GT4, £12,000 . . .

Add Subaru's three-year unlimited mileage warranty and acceptable fuel consumption (we achieved 28.3 mpg overall, and that involved almost exclusive hard use), and you have probably the best performance bargain around right now.

More than that, the Subaru Impreza is also a modern Japanese automobile in an extreme minority.

It has *character*. RRB

555
SUBARU
M54 KJW

Forest FLIER

Subaru's rally-bred Impreza is quicker yet £5000 cheaper than an Escort Cossie. So what's the catch? Peter McSean reports after 12,000 miles

12000 MILE TEST

I could see in the rear view mirror that the besuited bloke standing behind the Subaru wasn't about to move. Odd behaviour in a multi-storey car park, so I climbed out to see why. "Sorry," he said hesitantly, "but I just had to ask: is it true what they say about your car in the road tests? I'm thinking of getting one instead of a BMW 318i S, after Colin McRae won the RAC Rally... I was there, you know..."

Here was walking confirmation of precisely why we had chosen to put a four-wheel-drive, turbocharged Impreza on our fleet. Precisely, but not exclusively. Colin & Carlos Inc were going to give the car street cred by giving it streaked crud and that was an important factor in our choice, certainly, but there was an even better reason for running the hot Subaru. Over to you, Road Test Yearbook: "[The Impreza Turbo 2000 4wd is the] bargain of the year... quite possibly the performance buy of the decade."

Sure, it was cheap to buy at £17,499 when Escort Cossies were listed at £22,535, but would it also be cheap to run? Or would there be a hidden appetite for oil, tyres, hoses, turbos, brake pads or driveshafts? Would the areas where Subaru has cut costs prove too much of a compromise day in, day out? Could anyone live with such a dull interior, for instance? Would the car be a target for thieves? And would the tin box shake itself to pieces after ➧

For Point-to-point ability, easy-going in town, cheap, McRae drives one
Against Narrow boot opening, unsettled ride, mists up in the wet
Acquired 18 August 1994, new
What it cost then £17,543
Mileage covered 12,177
Breakdowns None
Warranty claims None
Repair costs None
Days off road Two (servicing)
Reports so far 7.9.94, 5.10.94, 16.11.94, 25.1.95

STAN PAPIOR

M54 KJW (above) had first taste of UK at Jarrow Slake, Tyne dock; 18 days later we drove it away from Holbeins

four months? Surely you only get what you pay for, and a car that covers 0-60mph in 5.5sec on its way to 137mph doesn't come that cheap unless there's a catch. Somewhere.

Well, if there is a catch we've yet to find it, 12,000 miles on. So far we've enjoyed the legendary Subaru reliability which gives the company so much repeat business. So far that bland cabin has mattered only to those who drive the car occasionally; frequent drivers soon don't notice it or find compensations elsewhere.

You don't have to look far. For a start there's the two-litre blown flat four, the performance of which improved markedly between 4000 and 8000 miles. Wielding 208bhp at 6000rpm and 201lb ft at 4800rpm, the 1213kg Subaru puts its power and torque down remarkably efficiently. Take its McLaren F1-equalling 0-30mph time of 1.8sec, or its 0-100 in 16.0, 30-70mph in 5.7, standing quarter mile in 14.3 (95mph). Then there's 50-70mph in fifth in 10.5sec, 30-50mph in fourth in 7.9sec. And for overtaking on country lanes, how about 50-70mph in third in 3.4sec? Or 30-50mph in second in 2.3sec?

Normally such acceleration is the preserve of supercars. Seldom is it offered by a car as versatile as an Impreza. In six months, our Subaru with the long name has been involved in two house moves (a fine workhorse with worthwhile space inside the car, provided you can access it through the narrow boot opening); transported five adults (one too many after about 30 minutes); acted as chase car on the RAC (a superb all-rounder); been the organiser's car for the London-Edinburgh Trial (reliability and speed proven again, this time in blizzard conditions); done more 200-mile midnight runs than I care to remember (where its easy cruising, lack of foibles and good seats and driving position mattered more than its choppy A-road/motorway ride); and trundled around and lived on London streets most days (where its Cobra three-circuit immobiliser has helped to keep it where we parked it). We've also admired its impeccable manners, gearing that lets it trickle along in traffic without feathering the clutch, its ability to pull out into small gaps without bogging down, good all-round visibility and 4340mm by 1690mm town-friendly dimensions.

Through it all, our super Sub has maintained a virtually faultless reliability record, needing no attention outside its scheduled services at 1000 miles (£34.78) and 7500 miles (£148.26). If those sums seem large, bear two things in mind: the only other outlay has been for petrol (a gallon every 24.2 miles on average), and the labour rate at our nearest Subaru dealer, Holbeins in London SW8, was a hefty £44.50 per hour. Though fairly typical for a London Subaru dealer, a quick ring round showed the £30 mark to be common for most parts of the country, with mid £30s in more densely populated areas. For top money you'd expect Holbein to be efficient, polite and collect/deliver the car during servicing, and that was the case.

So what's with the *virtually* faultless? Well, from delivery we felt the volume of air coming in through the ventilation system was inadequate. At the 7500-mile service Holbein gave it a thorough examination but could find no fault. However, the throughput of air seems to have increased, so we can only conclude that Holbein cured it inadvertently. Even so, the rear side windows and back screen still mist up easily in the wet and those who've forked out £1499 for the optional air conditioning say it's been worth every penny.

It's one of the small compromises I'm prepared to accept — though I don't think I should have to — in exchange for low-cost, high-thrill driving. There are others, such as the absence of low fuel and screen wash warning lights, the difficulty during daylight of seeing if the air recirculation button is lit or not, the lack of a light in the glovebox, central locking that works only from the driver's door and lacks a remote, bootlid hinges that intrude into the luggage area, the absence of a rheostat to alter the brightness of the instruments, and a reduction in front passenger legroom due to an ECU behind the bulkhead.

Spacious once past narrow opening

Brookland's test hill surrenders

Tight fit but basics easy to access

WHAT COLIN'S CROWD CAN DO FOR YOU

There is nothing wrong with how a standard Impreza Turbo handles. It's a tad soft at the front and understeers a bit in the wet if you're being picky, but that's about it. However, the folk behind Colin McRae's car, Prodrive (tel: 01295 273355), open up new levels of handling for road-going Impreza Turbo drivers. And they do so with Subaru's blessing. Stiffer, tuned suspension (£1322) means negligible body roll and minimal understeer, replacing the standard car's front-drive feel with greater stability, balance and poise. The stiffer front end also reduces weight transference under braking, avoiding the regular car's predilection for activating the anti-lock earlier than anticipated. The ride is hard, not helped by £1645 205/45 ZR17s, but we think keen drivers will accept the trade-off. Other wheel/tyre choices are on offer, as is a well-executed interior retrim at £1972. Our advice? Go the whole hog if you have the funds (£4929 all in). If not, start saving for the suspension mods.

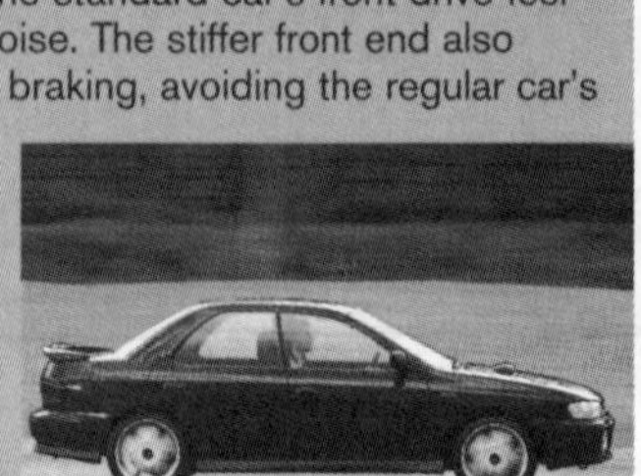

Prodrive's tweaks worthwhile

Surefooted on London-Edinburgh Trial in conditions that closed much of UK

This was it until odo read 1000 miles

And while I'm glad that the radio/cassette is removable by me rather than a thief, there have been plenty of occasions when I've found it inconvenient carting the whole unit around.

Yet I've appreciated the little things it gets right, such as being able to reach the manual aerial without moving my back off the driver's seat and the backlighting that makes the digital clock readable in sunlight. Then there's the way the headlights switch off when the ignition is off (parking lights are on a separate circuit), and the fact that there's a cradle on the fuel filler flap to hold the cap while you pour in the preferred choice of super unleaded (the higher octane level does make a difference to performance).

And we mustn't overlook the ease of entry and exit from any of the Impreza's seats, the logical placement of the controls in the cabin, one-shot driver's window, good-sized door pockets and glovebox, headlamp washers, well-shaped seats, driver's airbag (never used), superb main beam illumination (pity the dipped beams are such a disappointment), easily folded rear seats and a 0.60cu m capacity once done, 10.4-metre turning circle, three-year/ 60,000-mile warranty including Europe-wide breakdown recovery... the list goes on.

Above all, though, our Impreza impresses most for its point-to-point prowess, in all weathers. For the full story read our road test (6 April 1994), but for a compressed taster, read on. The one bar of boost means that you dictate your own velocity along twisting roads instead of finding your speed restricted by some bozzo towing a caravan or the two cars stuck behind it. While capable of improvement (see panel, left), the standard car's safe handling still entertains. Throw in strong brakes and an unerring ability to put the power down out of a corner long before most other cars could dare and there are few other cars capable of staying with the rally-bred Subaru, and none at this price.

Yet more astonishing still is how easy and economical it is to live with on a daily basis. Bargain of the year? After 12,000 miles, all the evidence indicates that our Road Test Yearbook conclusion is spot on — both for buying the car and for running it. Would we recommend this rocket for small pockets? Wholeheartedly. Believe it, Mr Suit in the car park, or it's your loss. ●

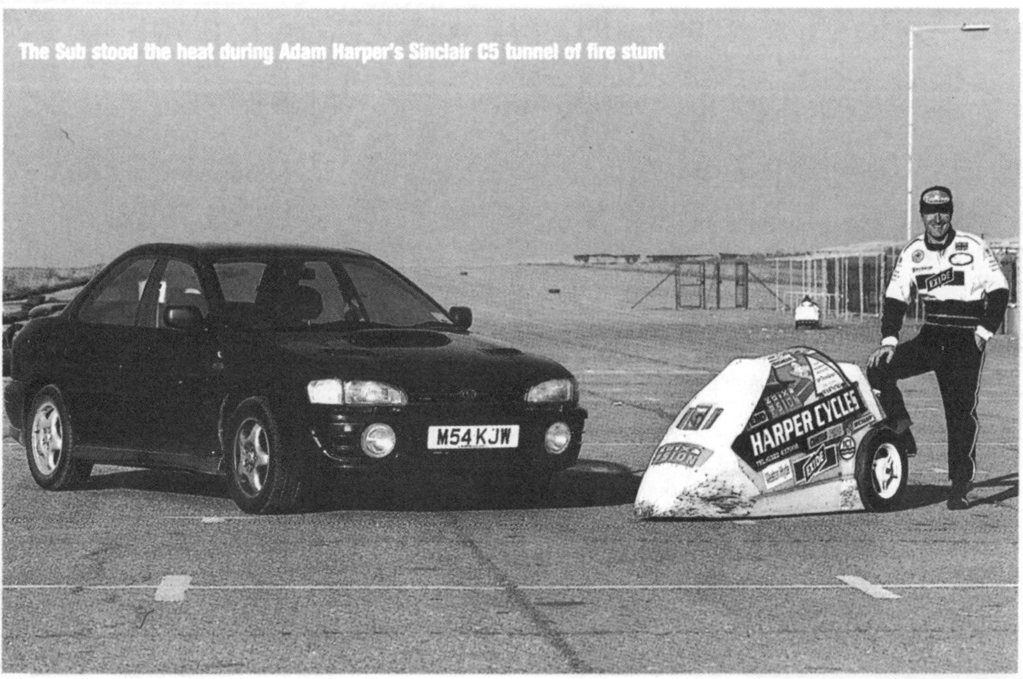

The Sub stood the heat during Adam Harper's Sinclair C5 tunnel of fire stunt

FACTFILE

SUBARU IMPREZA TURBO 2000 4wd

Mileage now 12,177
Date acquired 18 August 1994, new
Mileage when acquired 208
Colour flame red
Options floor mats (£44)

WHAT IT COST
Price new then, basic/with options £17,499/£17,543
Price new now, basic/with options £17,999/£17,543
Estimated resale value to dealer £12,500
Depreciation £5000/28.5 per cent

TYRES
205/55 ZR15 Michelin Pilot HX
Percentage worn front/rear at 12,177 miles 40/40 per cent (tyres swapped front to rear at 6985 miles)
Typical cost each £200

FUEL/OIL
Average fuel consumption 24.2mpg
Best/worst mpg 27.3/17.5
Litres of oil 0 (excluding services)

INSURANCE
Example 1 £601 (35-year-old married male with clean licence, five years' no-claims bonus, car garaged in low-risk Oxfordshire)
Example 2 £1366 (25-year-old single male with two speeding convictions, five years' no-claims and car not garaged in reasonably high risk Middlesex). Source: Quotel

ENGINE
Layout 4 cylinders horizontally opposed, 1994cc
Max power 208bhp at 6000rpm
Max torque 201lb ft at 4800rpm
Top speed 137mph*
0-60mph 5.5sec*
30-70mph through gears 5.7sec*
*Recorded at Millbrook Proving Ground with odometer reading 8107. Surface dry

DIMENSIONS
Length 4340mm (170.9in)
Width 1690mm (66.5in)
Height 1400mm (55.1in)
Wheelbase 2520mm (99.2in)
Kerb weight 1213kg

SERVICING
Mileage	Date	Cost	Labour/hour
1137	8.9.94	£34.78	foc
6985	16.11.94	£148.26	£44.50

Holbein Motors, 12 Ingate Place, London SW8 3NS. Tel: 0171 622 9003

REPAIRS AND FAULTS
None (but see text about the ventilation system)

SOLD BY
Subaru (UK) Ltd, Ryder Street, West Bromwich, West Midlands, B70 0EJ. Tel: 0121 522 2000

Subaru Impreza WRX

Meet the WRX, as in Rex the Wonderdog, as in the world's fastest mail-delivery wagon.

BY JOHN PHILLIPS

It's not the purple stripes that trigger the double takes. Not the purple competition seatbelts, either. It's mostly the steering wheel on the right-hand side that elicits the gasps. Lord knows, this Subaru Impreza wagon's squat, boxy shape otherwise attracts about as much attention as a Montana mongrel. Come to think of it, we began calling this rare model, officially dubbed the WRX, simply "Rex."

What with its mail-delivery steering and zero airbags and a unique turbo'd 2.0-liter engine, the WRX—a World Rally Championship homologation special—possesses no papers, no legal tags for operation in the United States. Subaru of America may soon resolve this issue, however. Here's why: In Japan, where this vehicle is legally sold (hence the right-hand drive), the WRX accounts for 36 percent of *all* Impreza sales.

The usual optional engine in the Impreza is an SOHC 2.2-liter flat four producing 135 hp and a gritty boxer growl. The WRX makes mush of that statistic, crashing through the 100-hp-per-liter barrier. Under the WRX's paper-thin aluminum hood (which shakes and wobbles like a Winston Cup car's) and lurking under a Hulk Hogan hood scoop that feeds an intercooler, there is nestled a 2.0-liter aluminum flat four treated to 10.6 pounds of boost. This arrangement is good for 217 hp at 6000 rpm, pouring out in a broad, flat, nearly seamless and eerily quiet rush to all four wheels. Indeed, the little turbo's kick is subtle. You feel some urgency at 4000 rpm, but otherwise you'll have to be told that this isn't a naturally aspirated engine. Of course, when you clobber 0-to-60 mph in 6.6 seconds (3.6 seconds quicker than the last Impreza wagon we tested), you pretty well know that *some* sleight of hand is afoot. Play out the WRX on its Wonderdog leash and it chews up the quarter-mile in 15.1 seconds at 90 mph—a terrier's hair better than a naturally aspirated Nissan 300ZX.

Top speed of 135 mph might have been greater were it not for the wind's view of an Impreza wagon, which is very similar to the wind's view of a Serbian bunker.

As it happens, Rex is even more

PHOTOGRAPHY BY JEFFREY G. RUSSELL

The right-hand steering and purple belts draw more attention than is strictly necessary. Competition seats are spectacular. Turbocharged flat four (below) produces 217 hp.

remarkable at *eliminating* wind. From 70 mph, we've observed stock Imprezas arrest all forward motion in 182 feet, but the WRX manages this same trick in an astounding 163 feet. It crabs to a halt with its ABS pounding and the tires scrabbling in a dance that tech director Frank Markus describes as "bordering on violence." Who cares? An Acura NSX can't stop this short.

The Krazy Glue grip is attributable to 10.9-inch front rotors, squat 205/55VR-15 Bridgestone Expedias, twin-piston calipers, a larger brake booster, and possibly 300 pounds of that gooey pine resin that Kirby Puckett slathers on his bats.

Speaking of resin, skidpad grip rises from the stock wagon's mundane 0.75 g to a stout 0.84 g, thanks to stiffer springs and dampers, trick alloy lower control arms in the front, new sway bars, and the aforementioned tires. Body roll is virtually eliminated at the expense of ride. If you've been trying to lose weight, expect to be reminded of your failure as each road pimple imposes belly-flesh jiggles.

Pushed hard on pavement, the WRX predictably understeers and goes all wormy at the nose. Pushed hard on gravel, however, the WRX reverts to its glorious rally roots. The driver begins to sound like a pitcher: "See this next hill? I'm hurlin' it in there hard, low and inside, a full-heat sinker."

When the WRX is on the marbles, the driver feels nearly no wheel slip, as tractive force is doled out flawlessly to whichever tire is most adept at accepting it. The result? You innocently dial up some truly horrifying speeds and ask your passenger: "Jeez, aren't those pine trees zipping past awfully fast? And *close*?"

This is as nimble a 2964-pound car as you will find this side of a Limbaugh-laden Miata, although the WRX (and possibly Rush, too) is almost 300 pounds flabbier than the standard Impreza wagon. Much of that thigh cheese can be blamed on the automatic transmission, expensively lifted from the Japanese domestic-version SVX. At the heart of this device is a rally-proven viscous limited-slip differential. During normal driving, torque is apportioned in a 50/50 ratio but can quickly skew to as much as 35/65, depending on how deeply into the dust you've dipped in your personal drive-home rendition of the Acropolis Rally.

Under full throttle, the transmission cracks off gear changes at an ambitious 6700 rpm. These are rifle-shot, blink-of-the-eye shifts that are, at the same time, surprisingly smooth. Downshifts are equally impressive. Punch the throttle at a 60-mph cruise and you instantaneously summon—sans lurch—not only a lower gear but also 5400 rpm on the tach, thigh-deep in the flat four's sweet spot. Very satisfying.

The only drawback is that this transmission binds and crabs and moans during tight turns and parking maneuvers—a noise from Rex we dubbed "park and bark." It reminds you regularly in the Wal-Mart lot that a snapped intermediate shaft to the center diff's first sun gear translates to a potentially ugly dialogue with an incredulous loan officer.

Many pieces on this car are definitely *not* attached to your Uncle Seth's Impreza. The purple four-point Schroth competition belts, for example, which look cool but are tedious to buckle and clash badly with flowered Hawaiian shirts. The competition seats are a happier revelation: perfectly contoured, with truly aggressive shoulder bolsters. Then there are the unique PIAA driving lights/foglamps. Illuminated to max yellow brightness, they stain the adjacent countryside until you'll swear you're driving through a vat of Vernors.

Rex the Wonderdog would do much to enhance the blue-rinse persona of Subaru's Impreza lineup here, never mind that only nine Americans have ever heard of the World Rally Championship or the WRX's heady place in the WRC firmament. All of this model's hard bits—brakes, turbo, suspension, transmission—already repose in the Subaru parts bin.

Of course, there are some automotive riddles to which a rally-prepped Subaru Impreza is not the answer, and no amount of taped-on purple shelfpaper stripes will overcome the trauma engendered by a $33,172 sticker. (Did we mention that Rex's transmission is, ah, *dear*? Did we mention what happened when American buyers faced a $30K Subaru called the SVX?) Nonetheless, if the question is, "What's the fastest way to deliver your Victoria's Secret catalog mid-blizzard?" the answer may well be this ingratiating little mongrel. We'll gladly assist with the adoption papers. •

Vehicle type: front-engine, four-wheel-drive, 5-passenger, 5-door wagon
Price as tested: $33,172 (Japan)
Engine type: turbocharged and intercooled DOHC 16-valve flat 4, aluminum block and heads, Subaru engine-control system with port fuel injection

Displacement	122 cu in, 1994cc
Power (SAE net)	217 bhp @ 6000 rpm
Torque (SAE net)	206 lb-ft @ 3500 rpm
Transmission	4-speed automatic with lockup torque converter
Wheelbase	99.2 in
Length	172.2 in
Curb weight	2964 lb
Zero to 60 mph	6.6 sec
Zero to 100 mph	19.5 sec
Zero to 130 mph	62.8 sec
Street start, 5 to 60 mph	7.7 sec
Standing 1/4-mile	15.1 sec @ 90 mph
Top speed (drag limited)	135 mph
Braking, 70–0 mph	163 ft
Roadholding, 300-ft-dia skidpad	0.84 g
C/D observed fuel economy	19 mpg

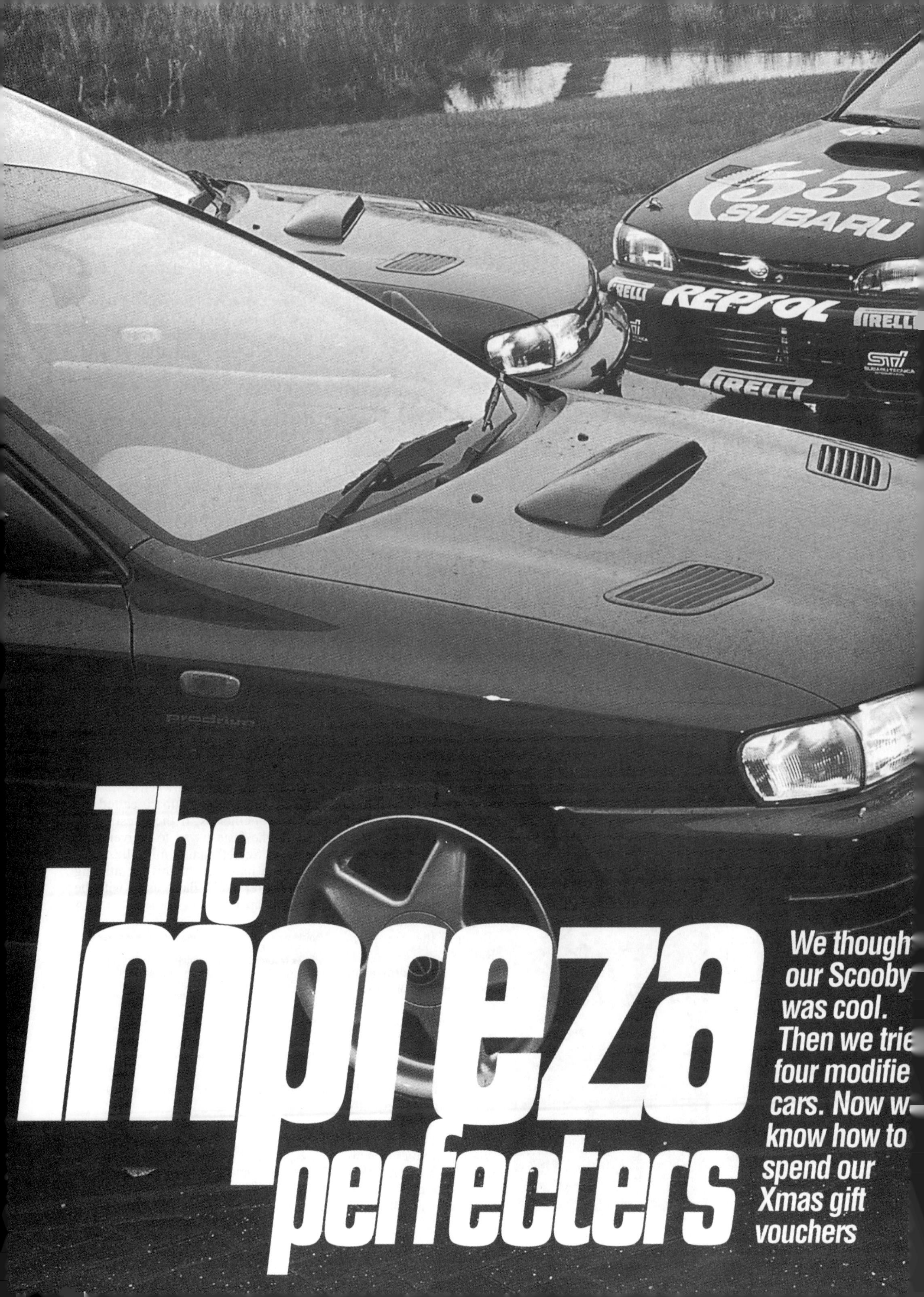

The Impreza perfecters

We though our Scooby was cool. Then we trie four modifie cars. Now w know how to spend our Xmas gift vouchers

1995 WORLD RALLY CHAMPIONS
1995 WORLD RALLY CHAMPION
555
REPSOL
NETWORK Q RALLY
PIRELLI
IHI TURBO
BILSTEIN
PIAA
NGK

WORDS RICHARD MEADEN
PHOTOGRAPHY DOMINIC FRASER

Forget Formula One – it don't mean a thing. Whinging prima donna drivers, racing that's tamer and more processional than a Friday evening trip around the M25, and cars that bear no relation to anything you or I are ever likely to drive. World Rallying, that's where it's at. The driving is absolutely, gob-smackingly awesome, Colin McRae is World Champion, and you can buy his car in a High Street showroom. Sorry Mr Schumacher, but I haven't seen many Benetton dealers in Peterborough.

OK, so you can't buy a Subaru Impreza exactly like McRae's but the showroom car is close enough. Or is it?

When we added the Scooby to our Fast Fleet we went for some lairy seats and a set of flash gold wheels to give it a bit of attitude, but some of you out there obviously feel we didn't go far enough. Here are four funky Imprezas that have gone a whole lot further.

NAME: Mike Wood
CAR: Prodrive Impreza Turbo
OCCUPATION: Projects Manager, Prodrive (01295 273355)

Prodrive Impreza

You know you've got a wild collection of modified Subarus when the car supplied by Prodrive, the people who made McRae's car, looks the most tame. Despite this, Mike Wood, Prodrive's sales manager, looks relaxed. His demo car is bloody good and he knows it.

As Mike explains 'Prodrive has developed a kit of parts that work together to do a certain job. Take away one part of the kit and the effect is ruined'. For example, if you try running the re-rated springs without Prodrive's re-rated dampers, the car will be worse than standard. Likewise, if you fit the suspension and use alloys with the wrong offset it'll be horrible.

Huge effort has been expended honing the kit to perfection, enhancing the Impreza without spoiling its competence or character. Attention to detail is clearly paramount at Prodrive. Rather than simply stiffen and lower the car for the sake of it, Prodrive has concentrated on the spring/damper rates and suspension geometry. The only stiffening has occurred at the back, with the fitment of a thicker anti-roll bar. The stance of the car is more aggressive but the aim wasn't to lower it for cosmetic reasons, hence the minimal 0.2in drop.

Wheel and tyre sizes and wheel offsets are also crucial to the car's set-up, the best compromise between style and function being 16in rims covered by 205/50 Pirelli PZero rubber, which is what this demo car runs on.

Get behind the wheel and not only does this Impreza feel infinitely sharper and more nimble, but the ride quality is more composed and less fidgety. Best of all, through a fast corner you can adjust the attitude of the car with the throttle. Running wide? Lift off, let the tail drift, feel the front grip and get back on the power. It's a delight to drive hard, which is probably why all the other Subarus here have the same suspension.

The Prodrive Recaro interior (as fitted to PC's Impreza) makes a massive difference to driver comfort and support, and to the overall ambience of the interior. If the cloth isn't flash enough then leather is available, as are carbonfibre or wood-effect dash kits and gear knobs.

For the full Prodrive effect you'll need to lavish £4700 on your Impreza, transforming an already exhilarating car into one of the quickest all-weather point-to-point cars you can drive. If this isn't enough, power boosting mods are on the cards. If you've got an Impreza you'd be a fool not to give Mike a call.

(Above) **The last thing you'll see before being blown into the weeds.** ***(Left)*** **Numbered Prodrive plaque on dash is the hallmark of a fully converted Impreza**

Barretts' Impreza

Kevin Croucher, from Subaru dealer Barretts of Canterbury, brought along an extra special Impreza. It started life as a 'Series McRae', the limited edition Impreza built to celebrate Col's win on the 1994 RAC. Barretts thought such a Scooby would be an ideal demo car and set about modifying it with Prodrive bits, Impreza WRX stuff and some bits of their own.

It's a lavish car alright, with rich blue paintwork and gloriously gold 17in alloys. Suspension is courtesy of Prodrive, as is the dash and switchgear trim kit. However, Barretts has taken the interior a stage further by fitting Japanese-spec WRX seats and trimming them with Conolly leather, along with the door panels, steering wheel and gear lever and handbrake gaiters. Perhaps the Series McRae wasn't the car to retrim, as it has the groovy Prodrive Recaro interior as standard, but the effect is opulent nonetheless.

(*Top*) Look at the wheels on that! (*Above*) Here are some more rare badges for all you Scooby spotters out there

NAME: Kevin Croucher
CAR: Series McRae Impreza
OCCUPATION: Subaru sales manager, Barretts of Canterbury (01227 766161)

The only other modification is the tailpipe which is fashioned from large bore stainless steel pipe and welded on in place of the standard car's twin pea-shooters. If you like the thrummy bass beat of the rally Imprezas then this mod is the best £70-worth you're ever likely spend.

There are plenty of people out there who might if Barretts' Impreza sales are anything to go by. According to Kevin, a third of all Imprezas they sell are Turbos - amazing considering it's the peformance flagship of the range. I wonder what percentage of sales the Escort Cosworth and Celica GT-Four account for?

To replicate the Barretts' 'Luxury Pack' interior you'll need to cough up £2300, while the wheels, tyres and suspension pack will add another £3400. If you fancy something a bit more hardcore than this, Barretts are the chaps to see; they run a Group N rally Impreza, so they know all the trick competition modifications. They'll also do their best to accommodate your wishes, however extreme. For proof of this you only need look at Richard Duncan's car, which they supplied and modified. ▷

"Series Michael" Impreza

Michael Edwards has very good taste. His list of cars-I-have-owned reads like something from your wildest lottery win dreams: Lotus Carlton, BMW M3, TVR 400SE and Chimaera, not to mention umpteen tweaked Cossies. And then there's his dark green Impreza Turbo.

Michael was turned onto the Subaru's charms when he heard about the Series McRae, but after looking at the car and its price tag he decided to buy a standard Turbo and do a 'Series Michael' for the same money. The fact that he's now blown his original budget is almost beside the point...

As you can see, Michael doesn't do things by halves. The wheels are the most extreme here, the interior re-trim is the most extensive, the motor's been chipped, and the boot is crammed with ICE. Oh yes, and there's the exhaust. Loud isn't the word. The general concensus is bleedin' loud. With a 3in bore, distinct lack of catalytic converter and two huge tail pipes, the sound of Michael's Impreza at large revs is awesome. Shut your eyes and you are Colin McRae.

You soon have them wide open again when his Scoob comes on boost though, for under the bonnet lurks a Superchips conversion that boosts power to close on 250bhp. Michael has been working closely with Superchips and runs development ECUs on his car, the current one being a 'hard' road chip. Roughly translated, this means the power comes in with a more brutal bang. As a result, it's a bit jerky on and off the throttle but the growing surge of power from just over 3000rpm all the way to the 7000rpm red line is addictive.

The interior is swathed in leather and suede, while the headlining is also trimmed in a suede-like material. Michael has plumped for Corbeau tront seats as they give similar race style and support to the Recaros but at a more reasonable cost. To add to the competition feel he's also fitted Prodrive's carbon-effect dashboard kit and gear knob.

According to Michael, all the aforementioned mods have been done without busting his original £23,000 Series McRae budget, but he's also fitted Prodrive's suspension package and an ear-bleedingly loud stereo system.

The modifications to Michael's car are very personal. He's a turbo driver from the 'nutter bastard' school, hooked on the top-end rush and incredible exhaust note. And he loves the aggressive, over-tyred look and the luxury of leather trim. The fact that the wheels aren't suited to Prodrive's geometry settings is an irrelevance to him. Michael sums up his feelings towards his Impreza thus: 'My M3 was just too civilised. I like a car that has humble origins, a few flaws and gives me a thrill. My only worry with the Impreza is that I can't sell it because there's nothing that could replace it'.

***(Above)* These wheels are absolutely enormous *(Left)* Big bore, no catalyst and it sounds like this "GRRRBWWRRRAAAARRPP!"**

NAME: Michael Edwards
CAR: Superchips Impreza Turbo
CONTACT: Speedwell Garage (01425 477788)

Richard Duncan's Impreza

You only have to look at the number plate on Richard Duncan's car to see what a stupendously sad man he is. Ex-Subaru works driver Carlos Sainz's famous reg is L555 REP, Richard's is N555 REP. Cool or what? Better still, the rally parallels don't stop there.

Richard, who also owns an Integrale Evo, bought his Impreza as a replacement for his last company car (a Carlos Sainz edition Celica GT-Four) from Barretts of Canterbury, who have been only to pleased to satisfy his rally fetish. Aside from having the *de rigeur* Prodrive suspension, 17in Speedline alloys and 215/40 Pirelli P700 tyres, his Impreza bristles with trick kit from Prodrive and Subaru Technica International (STI).

Scooby anoraks will be well aware that in Japan the Impreza Turbo comes in three different flavours. In addition to the 'cooking' Euro-spec Turbo, there are more powerful WRX and WRX RA models which have shorter gear ratios and increasingly Spartan trim levels. Think of them as Subaru's equivalent of Porsche's Club Sport and RS models.

Not surprisingly, Richard was gagging to put WRX bits on his car. Under the bonnet is a tough-looking STI alloy strut brace, while the doors sport gorgeous Group A carbonfibre mirrors. The standard front seats have been junked in favour of some fixed-position Recaro 'SPG' competition seats, just like the works rally cars, while the nasty airbagged steering wheel has also been slung out. In its place is a far nicer Momo item, which feels great but really shouldn't be there, should it Richard?

These changes are small potatoes compared to what's been done to the gearbox – Richard has had it filled with WRX ratios by the ever obliging Barretts. Fifth is now as low as fourth in the standard 'box, which limits the top end to 120mph but makes getting there a frantic, whining, gear-shifting frenzy. A nylon bush and block linkage gives the 'box a snicky feel. Carbon metallic brake pads bite the discs much harder than standard, while the back box from a Group N exhaust amplifies the Impreza's deeply throbby beat.

Richard's car is simply exquisite. It's extreme – too extreme for most tastes – and it's expensive (around £30,000-worth). But there is a method to his madness; when his Impreza has served its time as his company car he's going to do tarmac rallies in it, hence the gearbox, seats and suspension. For the time being though, he'll just have to put up with it as a road car. What a chore that'll be. Richard I hate you.

NAME: Richard Duncan
CAR: Impreza Turbo
OCCUPATION: Carlos Sainz impersonator

***(Above)* Alloy stut brace is cool, Momo wheel naughty but nice *(Left)* Look out, here comes Carlos!**

ROAD TEST

SUBARU IMPREZA 2,0 GT

A performance car for the individualist

COLIN AND Carlos, you lucky devils. You actually got PAID to drive one of these... At times, admittedly, it must be hard for world rally champion Colin McRae, who with fellow driver Carlos Sainz also earned Subaru the manufacturer's title in the competition version of the Impreza 2,0 GT. But this car is too much fun to keep a smile at bay for long.

The Impreza GT looks like a familar Japanese compact wearing some go-faster add-ons, has a name few but enthusiasts will recognise, and blows away most other things on wheels, irrespective of the quality of the road surface. No question, there's a certain charm to this nuggety little performer with the Jekyll and Hyde personality.

Two things help make Subarus distinctly different: flat-four engines and permanent all-wheel drive. This offshoot of Fuji Heavy Industries has doggedly followed its own distinctive drivetrain path, and it has garnered a similarly faithful following. What also helps is that in recent years the turbo 4wd Impreza has emerged as a major force in world rallying, since managing second place overall on its 1993 debut in the 1 000 Lakes.

Subaru says the turbo version is its best-selling saloon; a brief comparison of prices shows that abroad it competes against cars such as the BMW 320i, but here our import duty punts it rather higher up. Even so, its performance is out of all proportion to its price — or appearance.

At first sight the Impreza looks like a boy racer special, the unadventurous appearance of the standard model being given macho appeal by a bug-eyed look, courtesy of huge spotlights in the airdam, and a menacing bonnet scoop directing air to the induction system. Two smaller vents in the bonnet, on closer inspection, appear to serve little purpose - they are actually blanked off on the inside. Perhaps they're there for homologation purposes, being uncovered only in the heat of battle. At any rate, this ain't no boy's toy, considering it boasts a power to mass ratio (119 watts per kg) bettered by only a handful of cars — all much more expensive, and clearly sporting — in our current CAR Guide listings.

You won't find explosive response to the loud pedal in a higher gear, though; like a typical turbocar it lacks real low-down oomph. A glance at our overtaking acceleration figures will show quite clearly the initially ordinary performance, up to about 3 500 r/min. But wind it up though... with the boxer motor humming around 4 000 to 5 000 r/min in fourth gear, don't sneeze — even a toe-twitch will summon up Warp Speed.

Speaking of which, standing-start sprints are harsh (some say needlessly so) tests of a drive-train's ability to withstand the transition from immobility to maximum-effort forward motion. In the past, we have found with Subarus that engine speed had little apparent effect on the drive-train's reaction when we dumped the clutch: there would be an almighty "clonk" as the diffs took up the strain of all four wheels gripping, and the car would then streak off the line. So we crossed our fingers, wound it up, "... and that's 6,39 seconds," came the news from the passenger seat as we blasted past 100 km/h.

With the amount of torque developed

KEY FIGURES

Maximum speed	227 km/h
1 km sprint	27,06 secs
Fuel tank capacity	60 litres
Litres/100 km at 100	8,64
Estimated fuel range	496 km
*Fuel index	12,10
Engine revs per km	1 548
Odometer error	0,93 per cent under
National list price	R177 850

(*Consumption at 100 plus 40%)

Bug-eye spotlights and scoops lend a boy racer appearance to the front end. Opposite below, stability is good despite roll and understeer. Below, drab grey does nothing to lift the interior, though the Nardi steering wheel adds a nice touch. The standard airbag-equipped wheel is an option instead of the sunroof that came fitted to our test car.

by the turbo it was to be expected that similar harsh treatment would have the non-competition drivetrain approaching its design limits.

HOT CLUTCH

After only a handful of runs we noted a distinct smell of cooking friction materials and, after it became clear that the clutch was overheated and slipping, decided to call it quits. Had the road surface been slippery enough to give us a little wheelspin on takeoff, we suspect we may have done even better than the excellent times we did manage. With a bit more time and an unlimited supply of clutches we would probably have dipped under 27 seconds for the standing kilometre, too...

But what makes the Impreza really special is how it puts that power down, no matter where. On dirt roads, you can comfortably make like Colin or Carlos; it is quite tolerant of inexperience. On a gradual curve taken at speed it will begin to oversteer if you lift off, but a whiff of corrective opposite lock will enable you to hold it through the curve in a stable drift. With only 2,75 turns lock to lock the steering is really quick, although perhaps a little too much power assistance is provided. As the driving becomes smoother and speeds rise, the Impreza itself rises to the occasion. A series of really tight curves taken at modest speed begs for repeated action replays, the car effortlessly making the transition from understeer to tucking in to mild oversteer to drift. Even better, the Michelin tyres on our test car were more suited to tar than off-road conditions, which made them better for playing rally drivers at less than life-threatening speeds.

HANDLING

On the tar, in fact on the still-new Killarney tar oval, the car showed superb balance, with a bias towards power-on understeer, of course. At the end of the straight it was a sequence of brake hard, settle, turn in late, and then power outwards from the apex to clip the outside of the track. Under such conditions a good deal of body lean was noted, but this did not appear to affect stability much.

Choose your gear ratios properly, though: in lower gears the power comes in with a whallop and simply shoves the car outwards sooner than is ideal. For the adventurous, of course, there is always left-foot braking... The gearshift, incidentally, is light but lacks the positive, precise feel enthusiastic drivers demand.

As a performance car, it is clear that the GT merits great respect. Yet under the bonnet beats what has been described as only a mildly tweaked version of the standard Impreza engine.

The blower charge is cooled by an air-to-air intercooler, and fuel is dispensed by multipoint sequential injection. The

The short chopped-off deck, frameless windows curving into the roof and mild wedge profile add to the overall impression of solidity.

short-stroke 1994 cm³ engine, with alloy block and crankcase, pumps out 155 kW at 6 000 r/min and 270 N.m of peak torque at 4 800. This is transmitted to the wheels through a system of differentials. Up front the "centre" diff and viscous coupling split torque 50/50 front to rear under normal conditions, and at the rear a viscous limited-slip diff helps improve traction in cornering and on poor surfaces.

Unlike its normally aspirated siblings the turbo doesn't have a low "multiplier" ratio, perhaps for self-preservation if nothing else. When the turbo's spinning at full boost, there's such a thing as TOO much tractive force, as a few horribly stressed driveshafts and CV joints will confirm.

SUSPENSION

The suspension amounts to basically a strut at each corner, with a combination of trailing arm and lateral link providing wheel location at the rear. Subaru claim exceptional "lateral rigidity" thanks to what are said to be largest-in-class suspension struts.

Braking is to the same high standard as the rest of the car's performance. Subaru uses a tweaked version of ABS, said to give more natural pedal feel too. Speed sensors optimise braking in the "cross-piped" system by controlling the front wheels independently and the rears simultaneously. Supplementing the speed calculation function is a G-sensor operating with the viscous LSD, helping to avoid "cross-interference" between front and rear wheels. In the event, we recorded an average of 3,01 seconds in 10 stops from 100 to 0 — in the leading pack of ABS-equipped models.

Our steady-speed fuel consumption figures are more or less what can be expected from a two-litre motor having to overcome the extra drivetrain baggage carried by full-time 4wd. However, at 100 km/h — the figure on which our 12,10 litres/100 km Fuel Index figure is based — the engine is turning at only 2 600 r/min, well out of the power band. We have little doubt that the turbo's potential is likely to be unleashed quite often, with a sadly predictable effect on economy. Subaru claims urban fuel consumption of 11,4 litres/100 km.

Having used three-quarters of a tankful, we calculated that the Impreza had guzzled fuel at the rate of 18 litres/100km during our performance test, which included a couple of top-speed runs and half a dozen kilometre sprints. Incidentally, top speed was slightly below the claimed 230, with virtually identical speeds of 227 being recorded each way of our out-and-back test run.

With all this, the Subaru is surprisingly refined; its noise levels match those recorded by many of what are classed as luxury cars. However, that purposeful boxer-engine growl will be a constant and noticeable companion at cruising speed.

In South African spec, the Impreza GT

SPECIFICATIONS:

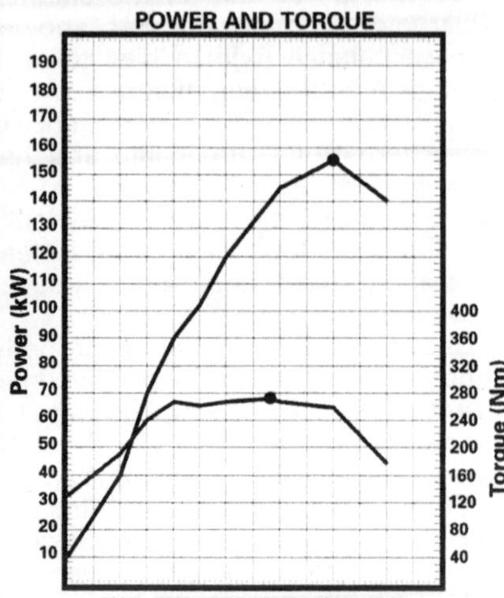

ENGINE
Cylinders .... four, horizontally opposed
Fuel supply .... multipoint sequential fuel injection, turbocharged and intercooled
Bore/stroke .... 92/75 mm
Cubic capacity .... 1 994 cm³
Compression ratio .... 8,0 to 1
Valve gear .... d-o-h-c, four valves per cylinder
Ignition .... electronic
Main bearings .... five
Fuel requirement .... 97-octane Coast, 93-octane Reef
Cooling .... liquid

ENGINE OUTPUT:
Max power I.S.O. (kW) 155
Power peak (r/min) 6 000
Max usable r/min 7 000
Max torque (N.m) 270
Torque peak (r/min) 4 800

TRANSMISSION:
Forward speeds .... five
Gearshift .... centre floor console
Low gear .... 3,454 to 1
2nd gear .... 1,947 to 1
3rd gear .... 1,366 to 1
4th gear .... 0,972 to 1
Top gear .... 0,738 to 1
Reverse gear .... 3,416 to 1
Final drive .... 3,900 to 1
Drive wheels .... 4 x 4

WHEELS AND TYRES:
Road wheels .... alloy
Rim width 6JJ
Tyre make .... Michelin Pilot HX MXV3A
Tyre size .... 205/55 R 15
Tyre pressures (front) .... 230 kPa
Tyre pressures (rear) .... 220 kPa

BRAKES:
Front .... ventilated discs
Rear .... discs
Hydraulics .... dual circuit, ABS
Boosting .... vacuum
Handbrake position .... between seats

STEERING:
Type .... rack and pinion, power-assisted
Lock to lock .... 2,75 turns
Turning circle .... 10,8 metres

MEASUREMENTS:
Length overall .... 4 340 mm
Width overall .... 1 690 mm
Height overall .... 1 400 mm
Wheelbase .... 2 520 mm
Front track .... 1 465 mm
Rear track .... 1 455 mm
Ground clearance .... 150 mm
Licensing mass .... 1 235 kg
Mass as tested .... 1 300 kg

SUSPENSION:
Front .... independent
Type .... MacPherson strut
Rear .... independent
Type .... dual link, struts, coil springs

CAPACITIES:
Seating 4/5
Fuel tank .... 60 litres
Luggage trunk .... 312 dm³
Utility space .... 960 dm³

WARRANTY:
2 years or 60 000 km

TEST CAR FROM:
Subaru South Africa

comes fully loaded. Standard features include air-conditioning, ABS brakes, either of driver's airbag or electric sunroof, radio/cassette player, and headlamp washers. Nice little touches inside include an elegant Nardi leather-bound steering wheel and gearknob. The driver's seat provides plenty of adjustment fore and aft and vertically, and most should be able to get comfortable behind the wheel.

After this extended song of praise it seems almost churlish to raise a negative: well, the Impreza doesn't feel particularly special from the driver's seat. The external (almost) Q-car looks are fine, but the shade of grey chosen for the interior does nothing to lift the bland ambience. Our test car came trimmed in cloth but buyers may choose leather instead.

TEST SUMMARY

Similar performance from other sources will cost you much, much more. That said, R177 850 is still a lot of outlay for performance you may not use most of the time. Then again, the Impreza GT is the kind of vehicle that encourages you to seek out places to enjoy its potential. It's practical, blindingly quick on any usable surface, and lots of fun; a rather different type of performance car for the individualist. ●

Peak power of 155 kW and peak torque of 270 N.m from two litres help propel the Impreza GT to a 27,06-second standing kilometre — but the superb traction of all-wheel drive plays a major role in putting that power down.

TEST RESULTS: SUBARU IMPREZA 2,0 GT

ACCELERATION

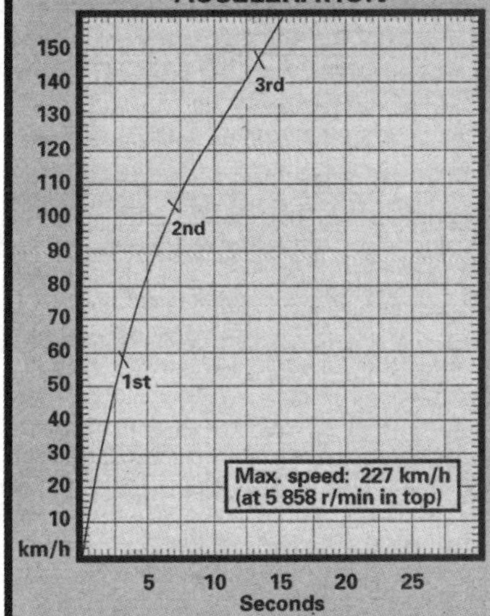

BRAKING DISTANCES

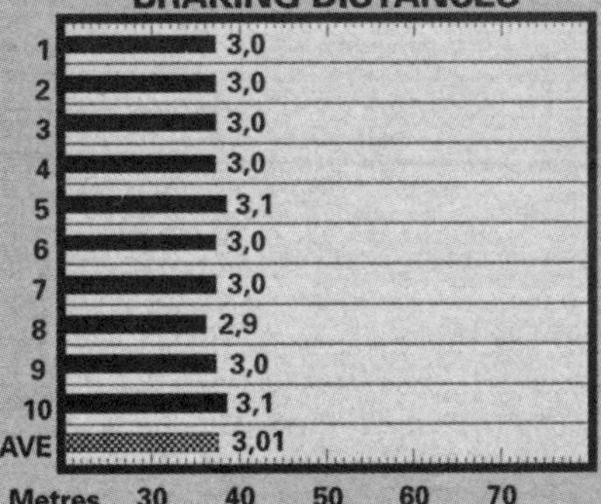

(10 stops from 100 km/h – individual stopping times given in seconds.)

GRADIENT ABILITY

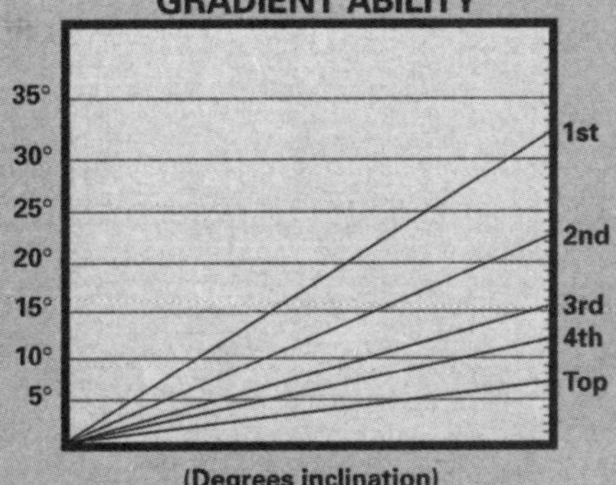

MAXIMUM SPEED (km/h):

True speed 227
Speedometer reading 237
(Average of runs both ways on a level road)

Calibration:	60	80	100	120
True speed:	58	76	96	117

ACCELERATION (seconds):

0-60 3,02
0-80 4,54
0-100 6,39
0-120 9,15
1 km sprint 27,06
Terminal speed .... 193,2 km/h

OVERTAKING ACCELERATION (seconds):

	3rd	4th	Top
40-60	3,33	6,18	9,51
60-80	2,39	5,08	9,00
80-100	2,34	3,76	8,22
100-120	2,51	3,62	6,07
120-140	2,96	4,12	6,08

FUEL CONSUMPTION (litres/100 km):

60 6,60
80 7,21
100 8,64
120 10,31

BRAKING TEST:

From 100 km/h

Best stop 2,9
Worst stop 3,1
Average 3,01
(Measured in seconds with stops from true speeds at 30-second intervals on a good bitumenised surface.)

GRADIENTS IN GEARS:

Low gear .... 1 in 1,6
2nd gear .... 1 in 2,4
3rd gear .... 1 in 3,6
4th gear .... 1 in 4,8
Top gear .... 1 in 8,3
(Tabulated from Tapley (x gravity) readings, car carrying test crew of two and standard test equipment.)

GEARED SPEEDS (km/h):

Low gear	50*	58
2nd gear	88*	103
3rd gear	126*	147
4th gear	177*	206
Top gear	233*	271

(Calculated at engine power peak* - 6 000 r/min and at max. usable r/min - 7 000 r/min.)

INTERIOR NOISE LEVELS:

	Mech	Road
Idling	43	-
60	60	-
80	62	68
100	64	70
120	67	72

(Measured in decibels, "A" weighting, averaging runs both ways on a level road: "mechanical" with car closed; "road" on a coarse road surface.)

ENGINE SPEED

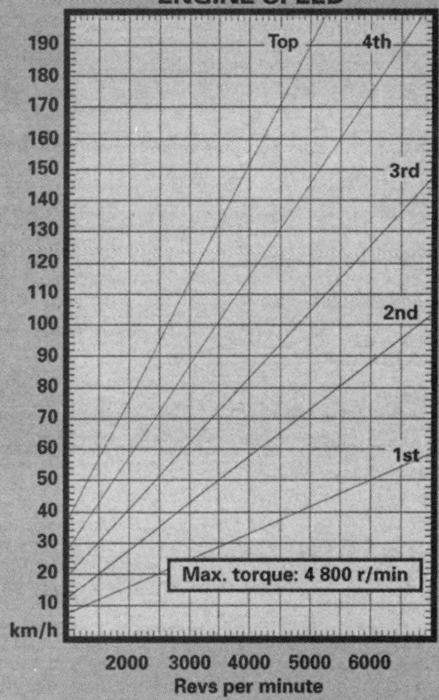

PERFORMANCE FACTORS:

Power/mass (W/kg) net 119
Frontal area (m²) 2,37
km/h per 1 000 r/min (top) 38,75
(Calculated on "mass as tested", gross frontal area, gearing and I.S.O. power output)

TEST CONDITIONS:

Altitude .... at sea level
Weather .... clear, warm, light breeze
Fuel used 97
Test car's odometer 988

CAR'S ROAD TESTS

CAR is the only South African motor magazine whose performance road tests are carried out at sea level – where naturally aspirated car engines achieve their (rated) peak performance – on a flat road to internationally comparable standards.

The test car carries a crew of two, a full tank of fuel and German Peiseler electronic road test equipment. CAR's fuel consumption testing is done with a Pierburg flow meter linked to the Peiseler computer. Top speed and overtaking run times are averaged both ways on a level road.

McRae's mean machine

SUBARU'S STUNNING NEW IMPREZA 555 WRC 97 IS THE FIRST OF THE WORLD RALLY CARS. KEITH OSWIN GETS EXCLUSIVE ACCESS TO COLIN McRAE'S 1997 WORLD TITLE CHALLENGER

'I want a car that looks like it's doing 100 miles an hour when it's sitting in the car park.' That was the instruction Prodrive managing director David Richards gave to his designers and that's exactly what he got. The first of the new generation World Rally Cars was shown to the public prior to the Catalunya Rally and Subaru's Impreza 555 WRC 97 finally ends the era of rally cars that look like something straight from the high street dealer showroom.

The new car is mean and aggressive; stunning styling from the pen of Peter Stevens – the aerodynamicist behind the McLaren F1 supercar – and every last ounce of technical wizardry that Prodrive's technical director David Lapworth can cram into it (see panel).

Outwardly different the Subaru may be but, in many ways, the World Rally Car regulations that come into force on January 1, 1997 are no more than a refinement of the current Group A regulations. Yet Subaru has become very excited about the new generation. Ford launched its own Escort variant a day after the Impreza, Toyota will have a new car by autumn 1997 and only Mitsubishi feels that it can stick to the current rules yet still remain competitive in the World Championship arena.

And Prodrive also believes that World Rally Cars will be the carrot to tempt more teams to the party, possibly doubling the level of support by the end of 1998. So what's so special about this new formula?

'The World Rally Cars are not that different, technically, from the current Group A cars,' says Richards. 'However the rules have been written in such a way that the homologation procedure we have to go through before we can go rallying has been made very much simpler. World Rally Car is not a new set of regulations, it's a different form of homologation process.'

Its purpose is simple. The cost of building and homologating a car suitable for World Championship rallying is phenomenal. Few manufacturers can justify the expense in the current economic climate of building 2500 versions of a car ▶

The most aggressive styling ever seen on a Subaru will take rallying into a new era

The new car may not be any longer than the old but it is certainly wider and therefore much more stable

◀ that, in its ultimate form, would be too expensive to sell, impossible to insure and a nightmare to drive in normal road conditions. No one, probably not even Subaru for very much longer, could justify the risk of having a car park full of vehicles that no one wanted to buy, just in order that a handful could be used to chase the World Rally Championship.

'I couldn't justify going to a manufacturer and asking them to commit to building a car like that,' says Richards. 'It takes a three or four year period to get a car from the planning stage of the road version to its debut on a rally. The new regulations give us a formula which allows any manufacturer to build a car competitive enough for the World Rally Championship. The turbocharger air restrictor puts a real upper limit on the power available and tyre width restrictions keep the speeds under control because of the limits they impose on handling.

'They also mean that manufacturers no longer have to do the difficult competition development work themselves. Any company like Prodrive, with an engineering capability, can go to any manufacturer and promise it a competitive car for World Championship rallying within 12 months. The new formula is crucial to the survival of the sport. It has taken away the long gestation period, opened up the market to a lot more manufacturers and reduced the cost by a factor of 20.'

> **We want people to look at these new cars, lust after them and want one for themselves**
>
> DAVID RICHARDS

Richards is, of course, a major supporter of the formula but is it only the 'in' crowd who want it? What about those on the outside? He insists: 'Those of us who are already on board would have every right to feel protective about the current Group A formula, because if you have a competitive car, it excludes the competition. But what is the point of competing unless there is someone to beat? That's why we needed a formula to guarantee future competition.'

So will there be an immediate rush to join the ranks? 'Not an immediate rush, no,' admits Richards. 'These things are never short term. You can't suddenly create a new set of regulations or a new formula and expect to see everone jump in and build cars the next day. Super Touring was the most recent success in motorsport, but it began six years ago.'

He believes that the new look of the cars will generate a great deal of excitement among people disillusioned by the 'ordinary' appearance of current machinery compared to the old Group B cars of a decade ago.

DIARY

February
Engine development begins in earnest although many aspects have already been evaluated. This development work will stretch throughout the building programme to the end of September. Concept design work starts and lasts for four weeks.

March
Peter Stevens begins work on exterior styling. This work encompasses the aerodynamic package plus the interior styling of the car. Styling takes 12 weeks and will involve the preparation of a full-scale mock-up which will be sent to the wind tunnel at MIRA (the Motor Industry's test facility near Nuneaton) for two weeks of final assessment.

April
Engine installation work starts, along with planning of the piping to the turbo intercooler. This process will last for 18 weeks. Design of the widened suspension also begins in April and this work, including work on the car's new cross-members, will last for 16 weeks.

May/June
Tooling for the body panels begins and will last for eight weeks. The rules allow for extensive modifications to the standard car with revised bumpers front and rear, wider wheel arches, larger transmission tunnels and bonnet cooling slots along with the rear wing.

July/August
Interior side impact protection is planned and tested, while items such as the wiring harness and dash panels are constructed. This takes around 10 weeks and coincides with the preparation and painting of the first bodyshell.

September
It only takes around four weeks to actually build the car itself, the time spent preparing and designing its components takes most of the seven-month process.

October
A nervous time for the Subaru team. Will it work? For two weeks at the end of the month a test team decamps to MIRA where the car is given its first shakedown runs by engineers David Lapworth and Tom Hunt. Colin McRae sees and drives the car for the first time on October 24 and gives it the thumbs up.

November
The Impreza 555 WRC 97 is shown to the media in Catalunya on Saturday November 2. Immediately after the rally the car is taken to a 'proper' test venue, where back-to-back tests are carried out with the 1996 car. This is the car's first genuine test in rally conditions.

'Previously we thought the fact that cars looked normal was a strength, but now we have come to understand it to have been a weakness. While it's right that rallying continues to use cars that people can relate to, it is crucial that we add to that an element of excitement, of glamour and of style. Everything we've done with the new car, from its overall design to its new colour scheme, has been geared to those elements.'

But style isn't everything. The car has to function effectively, if it is to be a winner. Which came first? 'I don't believe that function or style is capable of standing in its own right, but you can always add style to function. So that's the way round David and Peter looked at it.'

Richards has clear cut ideas about the way the sport should develop in the future, not only through the arrival of the World Rally Cars, but also through the events themselves.

'Rallying has, for far too long, been hidden away in the back of beyond with dirty cars and muddy forests. We need to get it into the public arena where we can build interest and create emotions. And you start with the product. We want people to look at these new cars and lust after them, to want to drive one. And they will be able to do just that.

'We're going to see a complete turnaround. In the past the homologation specials have been far too expensive and impractical. Now, with our new car there are maybe 50 elements that are unique to it from the aerodynamic package to the technical equipment. Each element can be taken out and put into a road car for the man in the street. And ▶

The 18-inch asphalt wheels with their Pirelli tyres sit snugly under the arches

The new Impreza (foreground) features a larger and more aggressive rear wing than the old

◀ that will be driven by market forces, rather than the necessity to build cars with all the bits right from the start. Motorsport should be at the leading edge, developing an image for car manufacturers and their sponsors. It shouldn't be the other way round.'

The advent of the new formula must sound warning bells for the lower orders. At the moment it seems unlikely that the new super-cars will eventually filter down to national competitions, and so should we worry about the long-term effects? Richards admits it is a valid point, but not a cause for concern.

'We should not look at the new formula as elitist,' he says. 'We should consider it opportunist. Today people can buy our cars and compete in local events. That means that a small club event can be easily won by someone who can afford an ex-Colin McRae Subaru Impreza, while the guy in his Vauxhall Nova has no chance. That isn't sensible, and the time is right to align certain cars to certain levels of competition.

'The World Rally Cars with the ultimate performance should be for the World Championship only, and we should consider that lesser level championships should be only for lesser level cars or World Rally Cars with lower levels of performance. The RAC MSA has seen the sense of this with the British Rally Championship and I admire them for making the decision to exclude all but the Formula 2 cars. We need to build a ladder of performance where a driver can start off in a lower performance car and gradually build to the top level. It's not so stupid if you consider that you wouldn't let a Formula 1 car compete in a Formula Ford race, which is exactly what ▶

We shouldn't look on the new formula as elitist. We should see it as opportunist

DAVID RICHARDS

INSIDE STORY

IF THE SUBARU LOOKS GOOD FROM THE OUTSIDE, WHAT'S IT LIKE TO SIT IN NICKY GRIST'S SEAT?

David Lapworth's words were hardly a source of comfort: 'If he's in the mood he's awesome round here.' I wondered if Colin McRae was in the mood as I strapped myself into the seat reserved for his new co-driver Nicky Grist. The belts were stiff, brand new and used by only a handful of people up to now.

So was the car. Barely run in and still to be shown to the public, I was getting a look inside this maiden's boudoir from the best seat in the house.

McRae reached down and pushed the big green starter button and immediately the boxer engine fired into life. But the familiar growl was gone and in its place a far crisper roar thanks to the revised exhaust system. If the old car sounded angry, this one really means business.

With a crunch the gears hit home and we were on our way, powering into what was, for me, the unknown. 'It's difficult to know what it will do around here,' admitted McRae. Well we were about to find out pretty smartish...

A long constant radius left hander led towards a rising 'fast left over crest' and the inside wheels lifted, the car trying to jump sideways. McRae held his line and the road straightened ahead of us into a vicious dip where I fully expected the sumpguard to take a wince-inducing smash into the ground.

But the suspension did its stuff and absorbed the impact and we were up and away again, heading for another, sharper right over crest.

The car rode this one better than the first, but appearances were deceptive and a sharp dip provided the launch pad for another airborne right hander before the road levelled and straightened again.

She goes! Colin McRae tries out the car...

...and his new high-tech instrument panel

Power on hard and we raced towards the second long left hander which led into a real test of the car's agility, a left and immediate right with something like 15-20 degrees opposite camber on each. My head rocked from side to side and I was glad I wasn't having to try and read the notes through here. McRae agreed it wasn't the kind of stage he'd want to drive every day of his life.

By now we were back into the first corner again, faster now than before, and with each subsequent lap the computer screen in front of McRae showed the improvement. Fourth time round and the tail was wagging wildly as we exited the final left hander, a sure sign that the abrasive surface had done another set of tyres. Wisely, we headed back to the garage, ready to stiffen the suspension and try again.

But McRae was happy and, if he wasn't in the mood right then, it wouldn't be too long before he was. 'It feels good right away,' he enthused. 'There's a lot more suspension travel and so the car doesn't get airborne in the same way as the old one. It's a lot wider and so that makes it very much more stable in the corners. The aerodynamics also mean it's not as nervous as the old car and I can go a lot deeper into corners without the tail getting really sideways. Although the engine isn't much different to the current one, its response is better.'

So he's happy? 'Oh yeah. I can't wait to get this car to Monte Carlo next January!'

1997 sees the end of the old (left) and in with the new (right) as rallying heralds a new dawn

Beneath the enlarged bonnet louvres lies Subaru's potent boxer engine

TECHNICAL MATTERS

PRODRIVE TECHNICAL DIRECTOR DAVID LAPWORTH TAKES US UNDER THE SKIN OF THE NEW SUPERCAR

L to R: Stevens, Richards and Lapworth

Subaru's new baby is based on the two-door Impreza Retna. What hidden secrets lie beneath its striking contours? David Lapworth will not be drawn on the amount of downforce provided by the aerodynamic package but explains: 'It's quite hard to quantify anyway,' he says. 'We haven't gone chasing massive amounts of downforce as it has to be a compromise over the whole year with events on different surfaces demanding different suspension settings and ride heights. But it gives a significant improvement over the current car in that we have turned measurable 'lift' into an equally measurable level of downforce.

'In the engine bay we can see several changes that have been possible with the new rules. Firstly we are allowed to fit an intercooler up to nine litres in size. This represents more than double the volume of the one on the current car and, as it is physically impossible to mount it between the inlet manifold and the bulkhead like the current one, we have moved it to the optimum location in front of the radiator.'

The intercooler is just one of several components whose dimensions are based on those used in the current Mitsubishi Lancer E3. The rear wing is another. 'When Mitsubishi made its Lancer E3 it was in a position to plan a car with the least compromise so the regulations have been written with that level of compromise in mind,' explains Lapworth.

'The inlet manifold has been turned through 180 degrees and the turbo has been realigned slightly. This allows the inlet pipe to the restrictor to be freed from routing under the inlet manifold, one of the problems on the current Impreza. Now we can fit a pipe that gives us an extra few per cent air flow, even though the air still has to pass through the 34mm restrictor. It probably only gives us an extra three or four horsepower but every little helps!

'We are designing a new transmission

The new wide arches being clay moulded

for the car,' says Lapworth. 'It will be able to use H-gate, semi-automatic or sequential systems. What you have to remember is that we designed the current transmission in 1989 and now we are keen to see what can be achieved with the latest technology.'

Stylist Peter Stevens designed a radical new dashboard layout. A glance past Colin McRae's steering wheel will reveal not the usual large rev-counter dial but a television screen where graphic imagery will flash up all the necessary information from the multiplicity of sensors in the modern rally car on 18 possible screen options. Eurosport is not one of them…

Prodrive is trying out different light systems, including gas-discharge which may eventually lead to the car being able to run without the light pods which affect the car's striking aerodynamics. Side-impact protection is built into the doors by way of lightweight honeycomb panels and a carbon-fibre panel which sits over the side-bars of the roll cage to keep stray elbows from going through the gaps.

Mechanics' safety is also paramount and the addition of a sealed fuel system goes a long way to protecting the spanner men who work on the car during refuelling. There is now an aircraft-type coupling linked to a pump which drains all the fuel from the tank and pumps back precisely the amount demanded by co-driver Nicky Grist.

'We've already seen enough to know that the new car is better and more stable than the current model. The first step is always the biggest and from then on any improvement tends to get smaller and smaller. There might be some big gaps between the teams at the start of the year but it won't be long before we're all fighting for seconds again and that's just how it should be.'

A basic rear wing on the first clay model

◀ rallying allows in club events.'

Richards feels the same should happen with the championships themselves. He feels that the World Championship should, quite clearly, be for the sport's top events and other series, even the Asia-Pacific, should be for entirely different rallies to further expand the ladder of opportunity. Providing there is no crossover of events, then World Rally Cars can be used on several series with different sized turbo-restrictors to maintain performance differences from World to Asia-Pacific and then to the European Championship.

'The new cars can excite the public and generate better television coverage, which has to be good for the sport. Event organisers still need to plan their event formats to make the most of this opportunity. World Championship rallies have to move away from being competitor-orientated to being media-orientated. They must make the sport more accessible to television and the media if it is to grow from being a minority sport into a global event with top quality rallies in all major continents. It won't happen overnight, but the process is in motion. 1997 is the first real test now that we've got rid of the rotation system, which served only to convince organisers that they cannot survive outside the World Championship.'

One of the most important steps forward for World Championship rallying is the arrival of this new breed of supercar, especially if it draws in fresh manufacturers at the top level. The new Subaru fulfils all the criteria and deserves to act as a catalyst for further developments.

With the Impreza 555 WRC 97, Subaru has redefined the shape of rallying in more ways than just its striking contours… ■

[LONG-TERM TEST] Four-wheel drive, 208bhp of turbo-a stage for the **Subaru Impreza Turbo**. But it's versatile

Subaru's hyperactive rally

charged thrust, heavy-duty fog lamps – all the world's and easy to live with too, as **Hilton Holloway** discovered

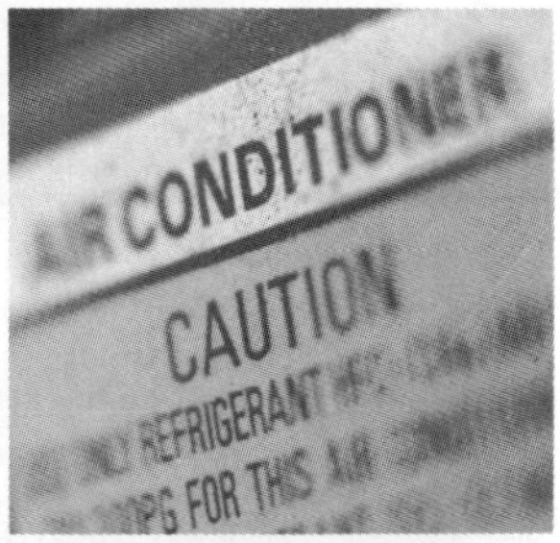

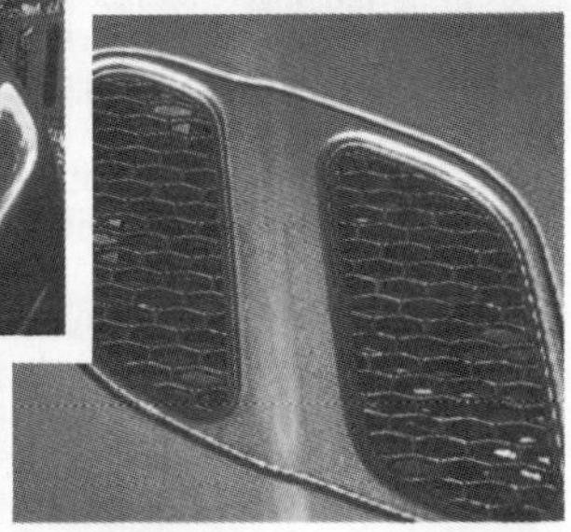

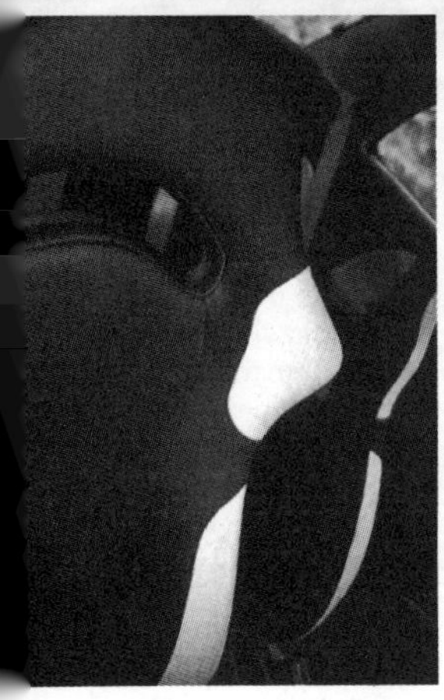

PHOTOGRAPHY BY **ALEX PUCZYNIEC**

IT'S A VERY SIMPLE RECIPE. FIT A medium-size car with four-wheel drive and a turbo'd motor and – eureka! – a road-going classic is born. Check out the roll call: Audi Quattro, Lancia Delta Integrale, and the Ford Sierra and Escort Cosworths. It's a winning formula: the convenience of a family car mixed with the power and grip of a racer, seasoned with the laurel garnish of success in motorsport.

As the outlandishly '80s Escort Cosworth limped to an early grave in the recessional '90s, Subaru quietly launched the Escort-sized, turbocharged, four-wheel-drive Impreza Turbo. According to our reader feedback, one of the biggest attractions of the Impreza was its relative anonymity. It could, many owners reckoned, be mistaken at a glance for any Far Eastern family saloon. This surprised us, until we realised that most of our readers were calling from the further-flung corners of the UK. City-based drivers like us were far more aware of envious glances.

Reputation and glowing road-tests were clearly the driving force behind the rise and rise of the Impreza Turbo. Most of the Impreza owners we talked to had read the magazine reports, taken a test drive and then opened their cheque-book. 'Oh, and my brother/son/friend has bought one as well...'

There's a certain clubbiness to ownership of this cult car. Fellow drivers often acknowledge each other, or even (whisper this) wave. But it's the combination of startling performance, nod-and-a-wink looks and clockwork running gear that unites Impreza Turbo owners in complete satisfaction. Only one of our readers wasn't keen on his Impreza Turbo, saying he much preferred his 150,000-mile Sierra 4wd estate.

champ takes it on the chin

ON DELIVERY

THE SUBARU arrived totally glitch-free, with just 167 miles on the clock. According to ex-features editor John Barker's notes (he was the original keeper), the Impreza arrived in 'absolutely flawless condition'. The paint was good, the panels fitted snugly, and the interior slotted together like a jigsaw puzzle. There wasn't a single rattle when it was new and there wasn't a single rattle when it left us six months and 9000 miles later.

However, there's no doubt it feels lightweight compared with the ubiquitous German competition. The doors close with a 'whump', but don't feel substantial. Lift the tinny boot-lid and you're faced with depressingly cheap boot-trim which doesn't even merit the description carpet. Barker's first impressions include the opinion that the car 'feels built for 100,000 miles, easy'. Yes, but not in the way a Mercedes-Benz does.

A standard-spec Turbo saloon costs £19,110 on the road and is only reasonably well-equipped. Four electric windows, central locking, a driver's airbag, alloys, ABS, electric mirrors, stereo, an immobiliser key, and that's your lot. There's no passenger airbag on offer, and although you can have a sunroof or air-conditioning, at £799 and £1549 respectively they're expensive.

Expensive or not, we were very grateful for the air-con fitted to our car, for in the summer the Subaru's stuffy cabin seemed to act like a giant storage heater, the dash soaking up the heat and then slowly releasing it. Two options we wished we'd had were the £327 CD player and the £1140 leather trim kit, which lifts cabin ambience onto another plane. There were also times when I would have swapped our metallic-blue saloon for the metallic-green hatchback that lived next door to me.

FIRST STAGE

'THIS CAR IS *soooooo* good. It shows surprising refinement. Integrity of the shell very impressive – it shrugs off pot-holes. The engine is a distant throb at the lights and it's quiet on the motorway.' Judging by his log-book observations, Barker was clearly impressed. Our '97-model-year car also boasted several subtle suspension mods. The work of Prodrive, which runs Subaru's rally team, they cured the earlier car's tendency to softness; our latest spec Impreza was also more refined, especially on the motorway, and gained impressively supportive rally-style high-back seats. Mind you, drivers under 6ft found them very uncomfortable – the prominent and non-adjustable lumbar support was the problem, pressing into the wrong part of their backs. Suited me fine, though.

GRAVEL RASH

UNUSUALLY FOR A modern car, the Subaru came with rigorous running-in instructions: for the first 1000 miles no more than 4000rpm, no sudden acceleration and no constant speeds, fast or slow.

Even before that milestone came up, however, the Impreza fell victim to a vicious and unprovoked attack. While the car was parked somewhere in Bedfordshire a nutcase came equipped with a screwdriver and caused serious damage to the paintwork. Both sides were riven with full-length scratches and the bonnet was deeply scored. The paint must have been thickly applied, however, as he'd (must

Keep your foot down, collect the bonus points from your friendly local magistrate

have been a bloke, surely) failed to penetrate the undercoat. It made for a deeply depressing sight, though.

With 1500 miles on the clock, it was safe to open the Impreza up. The essence of its excellence doesn't take long to pin down. No one particular aspect of the performance and handling stands out, but the car works together impressively as a harmonious and well-integrated whole.

Its prowess centres on the drivetrain, though: without permanent four-wheel drive, the Impreza wouldn't even be on the starting grid for greatness. It really is the only proper way to lay down 208bhp. As in an Audi, the engine is mounted longways, ahead of a longways-mounted gearbox. Unlike the Ford system (which had to convert rwd to 4wd) or the Integrale (front-wheel drive to 4wd) the purity of the Audi/Subaru layout means equal-length front driveshafts, a simpler gear linkage and a

Turbo'd flat-four dishes out 208bhp, has plenty of character

'The Impreza Turbo handles its power so well, it doesn't feel like 208bhp'

straight-through run for the propshaft to the rear wheels. The whole drivetrain is neat, evenly balanced and, above all, robust.

A nice contrast with the competence of the drivetrain is the oddball character of the flat-four engine. Although the spec sheet says the engine muscles in at 214lb ft at 4000rpm, it doesn't feel like it. The Impreza Turbo delivers plenty of horsepower, and searing performance (0-60mph in 6.5 seconds), but it doesn't have the heft of, say, the turbo'd five in the Audi S2.

Turbo lag was minimal: it seemed the turbo's wastegate was being carefully managed to allow the engine to build up steam evenly and progressively. If anything there was a hint of flatness from 2000-3500rpm, where the engine felt as if it was being denied the boost pressure it so eagerly wanted.

It wasn't many miles before I felt the car could do with another 40bhp, and many of *CAR*'s Impreza-driving readers agreed (and Prodrive, of course, will oblige here too). I decided this was simply that because the car handled the power so well, it sometimes didn't feel as dramatic as 208bhp should.

CLUB CLASS

THE EASE AND comfort of the Subaru made it especially good over long distances. It's almost impossible to set a value on the sense of safety and security instilled by the grip and stability – especially on poor roads and in the wet. You knew that whatever the weather, the Subaru would get you to your destination with time to spare.

My longest run was a trip to Aberdeen and back, the outward leg in torrential rain on a caravan-infested M6. The car's ability to flash past slower traffic without the slightest hint of instability or loss of traction was just fantastic. Four-wheel drive is a much underrated aid to long-distance comfort – even strong side winds and standing water fail to divert the Subaru from its arrow-straight course. Jump into an ordinary two-wheel-drive car and the difference is telling.

Although the Impreza's compact dimensions and speed make it a particularly satisfying commuting car, getting it out into the Scottish Highlands was a revelation. The sheer

wieldiness, the grip and wonderfully 'pointy' steering could not have been bettered on the empty B-roads. A car that can make assured and safe overtaking manoeuvres in such comfort and with such a compliant ride (the Impreza's ride quality was one of its biggest surprises) is an admirable engineering achievement. For one that costs less than £20,000, it's just amazing.

SERVICING

THE FIRST 'FREE' service came up at 1500 miles, when the car was due for an oil-change and basic check-over. Conveniently close to my south London home was Ingate Motors of Battersea, a Saab, Subaru and Isuzu specialist that said it could do the job without affecting the warranty. It ended up charging £41 (only labour is free with the 1500-mile service) but cleaned the car thoroughly – including the thick coating of brake dust on the front wheels. Ingate's boss must have twigged that the Subaru was a long-term-test car, so we were unable to take the car back there – annoying, for Subaru dealers, as I discovered, are spread thinly and often rurally.

By now, the Impreza had developed a couple of faults. Intensely irritating was a loud drone from the engine bay at 3250-3500rpm, exactly the portion of the rev-range used for motorway acceleration. Then the cassette-playing bit of the Philips stereo (dull-looking but fine-sounding) wobbled and died. Subaru's service guide told us there was a dealer in Kensington, central London. Handy. Less handy was Ace Kensington's service arm, inconveniently situated in Acton, west London. Dragging there and back was a two-and-a-half-hour trip.

They carried out the 7500-mile service early, but failed to fix the droning (which turned out to be an air bypass valve) or replace the cassette. Neither parts were in stock, but would be replaced under warranty. Ace said it would call when the parts arrived. We are still waiting.

LOG BOOK

DURATION OF TEST: six months
MILES COVERED: 9000
PRICE NOW: £20,879
PRICE THEN: £20,879
ENGINE: 1994cc 16-valve flat four
POWER: 208bhp at 5600rpm
TORQUE: 214lb ft at 4000rpm
OVERALL FUEL CONSUMPTION: 23.4mpg
MAXIMUM SPEED: 144mph
0-60MPH: 6.5sec
SERVICE COSTS: £154
OTHER GARAGE COSTS: £11.00 (puncture repair)
TAX: £169
INSURANCE: £972.09 (Norwich Union)
FUEL: £1082.05
OIL BETWEEN SERVICES: Nil
DEPRECIATION: £3509
COST PER MILE
EXC DEPRECIATION: 20.2p per mile
INC DEPRECIATION: 59.2p per mile

DRINK 'N' DRIVE

FUEL CONSUMPTION, and the car's effective range, became topics of debate. In town it could struggle to better 20mpg, and even dipped as low as 15mpg when the car did nothing but shunt around the capital. Although the figure improved as the engine freed off, it was much better in the country: an extended spell in Scotland saw it average 28mpg and crack 31mpg on one occasion. The fuel gauge was accurate, but even when on the red line the tank never swallowed more than 45 litres, effectively making it a 10-gallon tank, and giving the Impreza a maximum range of 300 miles. I became used to filling up when the trip meter approached 250 miles.

The Impreza didn't have a fuel warning light, like it didn't have a lot of other handy touches. It lacked map lights and central locking that worked from both front doors, and interior storage space was minimal – the '98-model-year Impreza addresses many of these criticisms.

This paucity, though, was forgiven by *CAR*'s Impreza-owning correspondents, one of whom said, 'It's a small price to pay. It shows they spent the money on the engineering.'

CONCLUSION

AS A MACHINE FOR fast and unfettered progress, there's little to touch the Subaru Impreza Turbo. It accommodates four adults in long-distance comfort and swallows their luggage, too, but it's also compact and relatively unobtrusive. It instils in the driver feelings of unencumbered freedom and security in equal measure, qualities that are usually mutually exclusive. There are few better driver's cars on the market. Considering the Impreza's devastating A-to-B performance, mechanical integrity and long warranty, this car is clearly one of *the* all-time road-car bargains.

All things to all men or nothing to no-one?

CAR READERS TELL US WHAT IT'S LIKE TO LIVE WITH THE SUBARU IMPREZA TURBO 4WD

	What did you have before?	Has it lived up to expectations?	What's good about it?	What's bad about it?	How about dealers and servicing?	Would you buy another?
PETER LEE, HERTFORDSHIRE SALOON, JULY 1997, 2500 MILES	An Escort Cosworth	Yes, definitely. I test-drove lots of cars in the £20k band, and it was easily the best	The performance, handling, practicality, looks, smoothness, build quality. It's a great cross-country car	The interior's a bit plasticky. Plus, it needs a boost gauge, more dials, and a sporty steering wheel	Quemby's of Baldock were very helpful. They supplied extras, including the Prodrive suspension	Yes. I'm looking at trading up to a Japanese-market 280bhp WRX version
MIKE KNIFTON, CLEVELAND HATCH, JUNE 1996, 25,000 MILES	A previously owned April '94 Impreza Turbo. I had nine Peugeot GTIs. A friend had an ultra-reliable Justy	It's exceeded them	The performance, ride, seats, A-to-B pace. And the comfort at high speed	Limited availability, waiting lists, money has clearly been saved inside	SG Petch of Richmond, Yorks have been very pleasant. Outstanding salesman, very friendly and trustworthy	I've already spoken to the dealer about a new model
DT TRAVERS, LEICESTERSHIRE, HATCH, JUNE 1994, 54,000 MILES	Renault 21 Savanna	Yes, it's been a joy to drive, it's versatile, and has excellent residual values	Value for money, low running costs considering the sophistication, the three-year warranty	Basically nothing. The interior is bland, but it's a reasonable price to pay	Sturgis of Leicester. Both the service and dealer have been excellent	I've just bought my second one
KEVIN HUNT, LANCASHIRE APRIL 1997, 14,000 MILES	I had a 25 grand budget, looked at a 20V Fiat Coupé, but wasn't taken seriously as a 25-year-old	Definitely. Thrilled to bits	It's a damn sight faster than I expected. Feels very safe, good seats. Extras include Prodrive suspension	You don't get much to the gallon, the brakes could be stronger	Gibson's of Brock. Fine. It's a family business where the MD sells the cars	Absolutely

LIE OF THE LAND

We've had a five-door **Subaru Impreza Turbo** on long-term **test** for a while now – that's the car that Colin McRae won his **World Rally Championship** in, and the car that **won** the manufacturers' title **again** this year. So we thought we'd **take** it to a few other **championships** from **Motorcycle Grand Prix** to **conkers**, **windsurfing** to **gurning** and **clay pigeon shooting** to the **RAC Rally** – plus the world **lying** title. The **truth** might be out there, but **we** went in search of some **prime porkies** instead

Longterm**test**

Little and Large are so funny that they require the services of a surgeon at every show to effect rapid repairs on ruptured flanks. Last year, Demis Roussos celebrated overtaking the Beatles in total album sales by hosting a cocktail party for 100 close friends inside his best smock. The Pope knocks about Vatican City on a Harley-Davidson. Vaseline tastes great on toast. And I'm hung like a donkey.

Now, we're all capable of such occasional porkies – witness a certain editor of my acquaintance, for instance: "In honour of Subaru winning the World Rally Championships, we've been taking our Impreza turbo long-termer to a selection of top World Championship events. Naturally, we've saved the most spectacular outing for you."

I rose, like a young salmon, to the bait and began the mental cramming of shades and suncream into my suitcase. Offshore powerboat racing in Italy? Skydiving in Portugal? Naked supermodel synchronised swimming in France? "Er, no, it's the World's Biggest Liar competition actually. In the Lake District. Wrap up warm now." Great.

There's no easy way to travel, traffic-free, that far north, but we shun obvious, odious motorway mayhem in favour of the breakfast-and-occasional-roundabout fun of the A1 and pause, mid-Little Chef sausage, to bone up on the background to the competition.

It began in the early 19th century, when a Cumbrian publican called Will Ritson achieved notoriety for talking consistent rubbish from behind his Wasdale Head bar. The valley in question was already littered with superlatives, boasting England's deepest lake (Wastwater), highest mountain (Scafell Pike) and smallest church (Wasdale Head). So, with Mr. Ritson's acclamation as the arch fibber of all time, the locals collected the full set.

Since then, the whole untruth issue has attracted enormous interest, and pilgrims from all over the world adjourn at the Bridge Inn at Santon Bridge – admission £4, including Tatie Pot supper – to have a crack at the title.

We're still picking the Chef's bacon out of our teeth when we come to a grinding halt for no apparent reason. We stroll past the massed ranks of static cars to have a look. A brand new shoe appears on the road. Then another. Then several thousand more. The lorry's on its side in the middle of the road and the driver is doing his best headless chicken act. He's clearly in shock because he's actually trying to pair up the shoes before frantically cramming them back into shredded boxes. Everyone stares open-mouthed, but no-one helps. We'll be here forever!

Ah, but there's a hugely tempting field to the left. The slope doesn't look too bad, there's a gap in the fence – and the Subaru's four-wheel-drive, isn't it?

The Subaru takes the slope with all the aplomb of the Jamaican four-man bobsleigh team and, as Phil struggles with a jammed camera, I keep the throttle wide open across a field wet enough to windsurf on.

Mud flies and the Impreza bucks, slithers and screams from lock to lock but refuses to bog down. Huge fun. Finally, we drive through an open gate and onto tarmac again. I want to go back for more, but Phil suggests, rather forcibly, that I shouldn't bother.

Now we're running late, so we hare off in search of a road big enough to feature on the map. "Oh bugger. What now?" growls Phil as, 20 minutes later, we hurtle towards another stationary queue. I leave him, banging his camera on the dash, to investigate.

It's a giraffe. No, in fact, it's two giraffes, crated up on a low-loader. And a railway bridge. The hapless lorry driver explains that, when scared, a giraffe comes over like the opposite of an ostrich and stands bolt upright. Which would be all well and good were it not for the fact that he's persuaded one beast to bend down for the bridge, and got halfway through, only to discover that the other is having none of it. And now, with both trembling beasts locked in the vertical, he's never going to make it to nearby Woburn Safari Park. Or anywhere.

"Acacia," he mutters. Sorry? "Acacia. They eat it. It's the only way I'm going to able to persuade Phydeaux to bend down," he moans.

Where can you get some, then? "Africa," he sobs. "But they've probably got some at Woburn."

"Work, you sod," Phil yells as I report back, which I think is a bit rich. "Not you," he snarls, "the camera." Still no joy, then. I explain that I'm going to drive to Woburn to collect some giraffe food, in the hope that he's sorted out things in the interim…

Zookeeper Martin Honeywell is in a right old lather. He's been expecting Phyliss and Phydeaux for hours. "Not

CLAY PIGEON

August 10/11 saw this fella, Gary Phillips, British clay pigeon shooting team captain and former World Champion, have a crack at the Express World Sporting Championship in Shropshire. Sadly Gary didn't win, but fellow Brit Carl Bloxham did, by hitting 203 clays and seeing off 550 rivals from 16 nations **AS**

BIKE GP

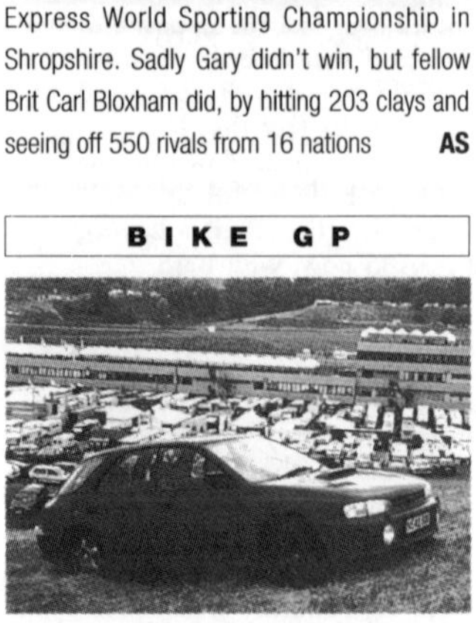

At the end of May, Art Maestro Marcel and I blasted down to Mugello in mid-Italy to catch the motorbike GP. The little Impreza did a manly job of tackling miles of autoroute and autopista, also hacking across the Alps at high speed. Our only niggles were the rather soggy seats and the lack of air conditioning. It's even the right colour for Italy… **CM**

again…" he mutters. With the back of the car brimming with three-inch-long thorns, Martin and I bleed our way back to the bridge. Mercifully, it works. Sadly, Phil's camera still doesn't.

Finally, we get to spend some uninterrupted time on board the Subaru and take stock of our surroundings. It's comfortable enough but rather drab, like a ballgown fashioned out of a binliner. The answer, of course, is to opt for the sporty interior upgrade; far more appropriate for such a fast car.

And the Impreza is indeed seriously fast. Subaru have turbocharged the nuts off their growly old 1,994cc, flat four unit, and it's good for a respectable 208bhp. This equates to 60mph in just 6.2, and a full floor-mat molestation of 138.3mph. Roughly.

But it's not an easy car to drive smoothly. The clutch is wicked, giving out altogether at 17,000 miles and slipping whenever the turbo comes on song in the gear of your choice. Subaru fixed it for us, but it still combines with an eager turbo to make progress in a straight line a series of over-enthusiastic, interconnected lunges.

On top of this, we must add a somewhat agricultural gearchange. Accelerate hard in first and the lever is so eager to get into second that the slightest touch sends it leaping down into neutral. Same for third to fourth. And the springing to neutral is too weak, making fifth to fourth a slower process than it should be. Furthermore, at speed in fifth, the box whines like a fiscally-challenged child in a sweet shop.

Our squirt up the A1's a high-speed hoot, the A66 an even more fine drive bordered by snow and terrorised by the RAF, whose Tornados rip overhead at zero feet. The first time it happens, I drop my jelly-baby in shock. Bit of a bugger, it was a black one, and there's no chance of finding it in the Subaru's sombre footwells. But the second time, it's the Tornado that drops something.

We watch, gobsmacked, as a black blob with nasty little fins falls lazily from the underside of the plane.

"Bloody hell. That looks like a b…" Phil never finishes the sentence, because the offending item has just introduced itself to the cab of a passing Rumbelows lorry at a speed of around 400mph. We hide. But there's no 'BANG!' Reg

The Subaru Impreza Turbo can cope with most things you throw at it, from three-foot high killer mallards (there's a secret farm in the Lake District, you know, where the Army breeds them as vicious attack ducks) to crowds of northerners applauding more magnificent mendaciousness

Thumb, the driver, is the most phlegmatic man I've ever met. Totally unmoved, he stands peering through a neat five-inch hole in the passenger side of his windscreen. By now, the Tornado's back, circling guiltily. Reg shakes everything at it and wins the World Swearing Championships hands down as he flings open the tailgate to check his load. Nothing. Not a sausage. Just loads of cardboard boxes.

"What the bloody hell am I supposed to do now?" he snarls. We helpfully suggest he might drive to the nearest bomb disposal depot. "Not f***ing likely, mate!" he bellows. "It might f***ing go f***ing off!"

Phil, whose camera is still not responding to being sworn at, is by now so despondent that he couldn't care one way or the other. So we leave Reg waiting for the bomb squad and pray for an incident-free, if wet, last leg towards the lakes.

Those scoops on the bonnet don't actually help airflow, or cooling for that matter. No, they're designed so that during a postal strike red Imprezas can be parked on every corner and used as substitute pillarboxes. That's what this bloke would probably have told you anyway

The Impreza oozes high-speed confidence on these drenched road surfaces, giving the impression that it's not going to budge an inch out of line, no matter what lunacy you conjure up.

We tool into the Lake District to find that the steep bits have received an icing sugar dusting from the great *sous-chef* in the sky. The place is awash with anoraks of a primary colour, hiking, mountain biking and even kayaking on the lakes. All, to a man, have entered the World's Reddest Nose contest, *and* the World's Stupidest Beard contest. The good news is that this leaves the roads relatively clear of old folks taking the travel rug and parcel-shelf tissues for a spin. The other good news is that Phil's finally coaxed some life out of his camera, and I spend a cosy afternoon entering him for the World's Coldest Photographer contest by pretending not to understand his distant, desperate, car positioning signals.

Arriving at the Bridge Inn, we find the place already heaving with many fine examples of last year's Christmas jumper and the huge pub laugh. Proceedings get under way with a terrifying, mass-karaoke mumble through some song or other. I, for one, can't understand a word anyone either speaks or sings of the local dialect. Instead, I peruse the rules of today's contest, which are simple: No mechanical aids

RAC RALLY

"Welcome to Radio Five Live's Network Q RAC Rally Report. The remaining three cars in the event, all works Skodas, have been ordered to hold their positions and drive sensibly." It was a dull RAC, but at least the Subaru looked the part and went like stink. Maybe we should have entered. We might have won **AF**

WINDSURFING

Meet Scott Fenton *(left)* and Nik Baker. They were supposed to be competing at the British round of the PWA Windsurf World Cup in Brighton in September, but there was no wind when we went and so no racing. Kiwi Scott is a "wild man from the land of the long white cloud" and Nik is the best Brit in the world, with indoor wins in Amsterdam and Paris **MH**

C O N K E R S

On the weekend of October 12, we had two options: to go to the San Remo Rally or send mad Mexican photographer Tony Olmos to the World Conker Championships in Ashton, Cambs instead. Tony took rather a shine to the sombrero-wearing competitors from Portsmouth University (above) and missed out on much of the action, which saw 450 conkerers doing their thing. The organisers supply the 3,000 conkers required for the competition and string them all on regulation leather thongs, to prevent any schoolboy-style pickling or dodgy knots. There are two main techniques – the traditional downwards 'donk' or the sideways 'slash' favoured by many of the younger, flashier contenders. But in the end, it was the experience of 74-year-old John Bull and his donk that won the day **NB**

permitted. Dialects permitted. Any overseas competitors not speaking English must have an interpreter. Members of the legal profession and politicians are barred from entry.

The 'Tatie Pot' is accompanied by a red cabbage so viciously pickled that the full 'dragon's nostril' is a racing certainty for tomorrow morning. Meanwhile, the judges – the Mayor and Mayoress of Copeland, assorted burghers, the editor of the *Whitehaven News*, and Mr and Mrs Jennings Brewery – jostle for position as the first of nine contestants takes the stand.

First out of the traps is Mr John Reeves, a 21-year-old 'rabbit angler' who didn't enter last year but told the *Border News* that he'd come third. So they printed it. A fine start. Sadly, his tale of Uncle Hercules who ran the Bacon Heritage Centre and invented the Cumberland sausage which, due to resultant wind, caused the extinction of the dinosaurs, rather loses its way. Better than the efforts of 'hot cross bun designer' Taffy Thomas, however. He simply tells a string of amusing jokes, without a single, home-grown lie.

Our third contestant, Derek Martin, has written a cookery book called *50 Things To Do With A Rhino*, which starts with the encouragingly daft "first pluck your rhino..." But it's all downhill from there, and there are dark mutterings from the throng concerning the quality of this year's entries.

The fourth entry is none other than the BBC's own John Walters, who makes a brave stab at explaining that cost cuts have forced all well-known BBC figures to be replaced by actors, that Sue Lawley is actually two men in a skin, and that he himself did 18 months as her back legs.

But the next contestant brings the house down. His name is John Graham and he's a local guinea pig farmer from Longcummercattif (honest) with hands

THE TRUTH	
Price	£19,419
Engine	1,994cc flat-four turbo, 208bhp
Performance	0-60mph 6.2 secs, 138mph
Test/EC combined mpg	26.2/27.2
Mileage to date	19,607
Faults	loose heat shield, juddery brakes, premature clutch wear
Costs	£154.02 for two services

The interior of the Subaru is bright, colourful, and its well-padded, supportive leather seats complement the carbon fibre trim and high-quality plastic... Actually, that is one lie even these blokes probably wouldn't applaud. The drab, flimsy-feeling interior was its worst point

like shovels and an accent so thick you couldn't pour it off a spoon. John has won for the last two years running.

He proceeds to tell an astonishingly funny tale concerning a turkey. Well, the few words I understand are funny anyway. It seems that, one year, his wife brought home a baby bird to rear for Christmas and it became the family pet. When fully grown, it went shopping with her one day but disgraced itself by attacking the poster of the local Labour candidate in the butcher's shop window and causing major damage. So she left it outside the shop, where it got a parking ticket... Oh, you had to be there.

The next entry, John Storey, apparently told a wonderful tale last year about an exhibition quality turd that he took round in a shoe box, but fails to match it this time. Reg White took rhino horn aphrodisiac on his honeymoon and spent the night ramming a Land Rover. Cliff Atkinson disguises a dig at the size of vet's bills beneath a rambling piece, which causes a stir of assent in the by now drunk, rowdy and disinterested audience. And George Summer from Macclesfield talks a load about woad. This, it transpires, is the furthest anyone has travelled to compete.

The judges retire and, as expected, John Graham collects a third consecutive Biggest Liar in the World tie. But he'll be sitting it out next year, in accordance with the rules.

Typically, with camera poised for action, the journey home is uneventful. But there is something about the Subaru that keeps you amused behind the wheel. It's stylistically all over the place, raw in every respect and, with the turbo on song, a blast to drive. Best of all, whenever we take it for a spin, life always seems to end up almost too good to be true. We'll miss it.

Shame about that camera, though; no one's going to believe a word... □

With the back seats down, the boot has an area of over 47 hectares, making it just large enough to park a short-wheelbase jumbo jet or stage a rugby sevens tournament in. And the Impreza's wheels are rather nice alloys and, oh, about this wide. Did anyone see an emu?

OUR VERDICT

We'd buy one because of its brilliant engine offering extreme performance, its excellent handling and reasonable price, general reliability and high standards of customer care at the dealer

We wouldn't buy one because the clutch action is a bit tricky, making smooth town driving difficult and probably contributing to its early demise. The interior is deeply dull and the seats need more support. Other than that it's just plain fab

GURNING

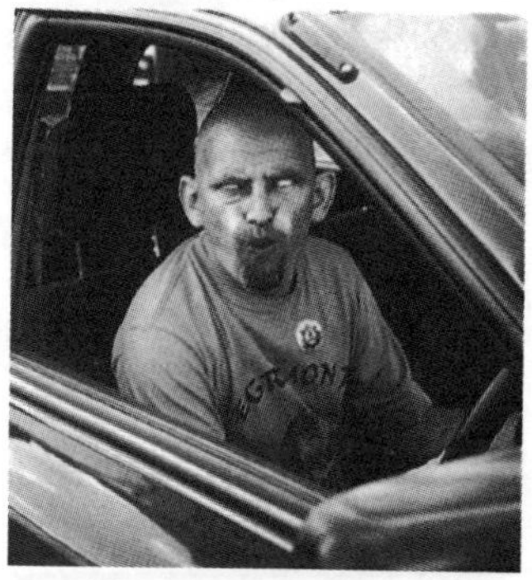

This man is not grimacing at the true horridness of our Subaru's interior, he's gurning for joy at having just become the World Gurning Champion. In September, we took the Subaru to the world's weirdest championships, all the way up in Egremont, Cumbria, not too far from sunny Sellafield. Victor and local hero Peter Jackman *(the gargoyle in our Impreza, above)* took time to talk us through the rules: "We're judged on the grotesqueness of our grimaces and the extent to which our features change. Dentures can be removed or reversed and you can heighten the dramatic effect by thrashing around on stage and making wild, animal-like noises." The result was truly horrific, but none were quite as stomach-turningly horrible as Australian Jim Wilshire *(right)*, doing his rabid kangaroo impersonation. Utterly disgusting, but very funny **PG**

VISCOUS COUPLING

LANCIA'S INTEGRALE, LONG-REIGNING KING OF THE RALLY DEVELOPED ROAD CAR, FACES ITS STIFFEST CHALLENGER YET – THE SUBARU IMPREZA TURBO 2000. ANDREW FRANKEL REPORTS

photography by Stan Papior

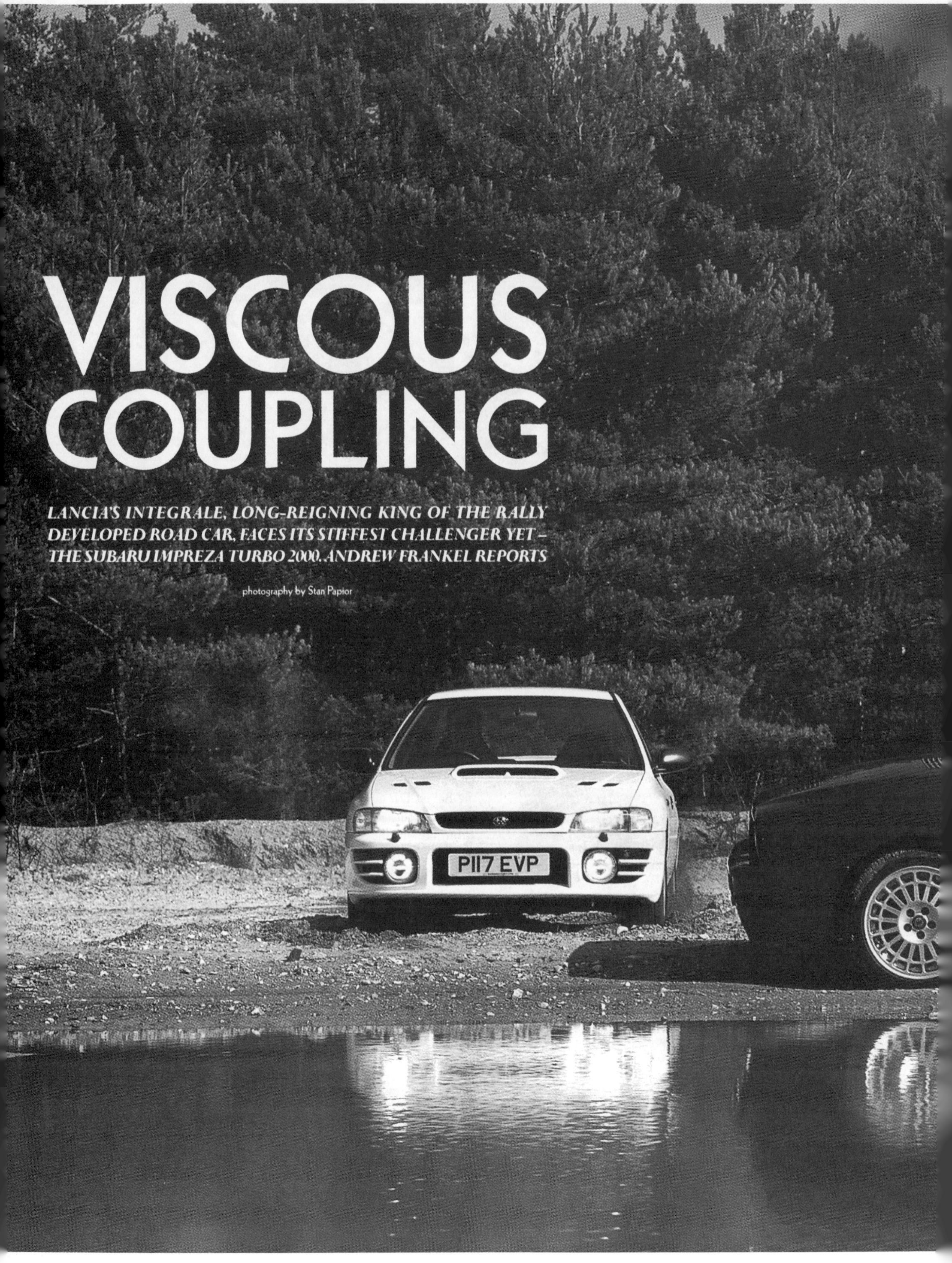

Integrale carries a great deal more weight than the Impreza, but you wouldn't guess it from the way it handles. Steering is pin-sharp and accurate

THE PEOPLE AT PRODRIVE, THE Banbury-based organisation that prepares and runs Subaru's World Rally cars, really wanted to win the Portuguese rally. In itself, there is nothing unusual about this; Prodrive, of them all, was not there just to make up the numbers. But Portugal was different; if either of its Subaru Imprezas had been first across the line, Subaru would have equalled the Lancia Integrale's record of six straight championship wins and, with it, established a credible claim to being one of the two most successful rally cars of all time. In the event, it never came to this as, in a rally of unusual mechanical destruction, neither Impreza would even see the finishing line, much less be first to cross it.

Yet even now, in the midst of Subaru's domination, it seems hard at first to square a marque of such traditionally humble and functional aspirations with the rallying greats such as Saab, Audi and, above all, Lancia. We should not be so patronising. Subaru has not been content, like Skoda, to leave its giant-killing to the special stage. It is after the legendary status of the Integrale on the public road too.

The Subaru Impreza Turbo 2000 seems an unlikely device to replace the Integrale in the affections of those who, throughout its years at the top of the world rally rankings, fell in love with the little Lancia and bought it despite the limitations of its left-hand drive only layout.

Place an Integrale – any will do, but the last, Evolution 3 model, is the best – next to an Impreza and the Subaru looks more than faintly ridiculous. It seems so tall as to be on tip-toes, its humdrum tin box styling curiously at odds with its dramatic bonnet scoops and louvres. Pretty it is not. The inside hardly helps Certainly the steering wheel is on th correct side of the car, which is some thing, but this does nothing to offse the murky cabin with its dark gre expanses, plain switchgear and du instruments. Here you will find noth ing, save an unconvincing hole in eac of the front seatbacks to acccomodat a race harness, to suggest that this i anything more than one more anony mous, four-door, family slogger.

The Integrale could not be mor different. There is not another car i the world whose essential shap would benefit so much from th

Subaru has the legs on the Lancia on the straight and smooth stuff, but though it handles well it lags behind when the corners come thick and fast

umped up addenda that surround he Lancia. Spats, slats, scoops and vings cloud the lines Giugiaro penned or the Delta almost to the point of omplete obscurity, but where you night expect the results of such basardisation to be a caricature, it is, in act a marvel. Low and squat on its pecial Bilstein struts, this Integrale xudes purpose like few road cars at ny price. Stop anywhere for even a hort period of time and passers by vill stop. And they will stare.

The inside, frankly, is a mess but till a wonderfully enigmatic mess. Vhere the Subaru makes do with minimal instrumentation, the Lancia provides dials for oil pressure, oil temperature, volts and turbo-boost pressure, doubling the number of gauges in the Subaru. Here too you'll find such dubious delights as electric window switches by the handbrake, three column stalks, and a clock where the roof meets the windscreen.

Yet the things that matter, as is so often the case with cars of such Italian eccentricity, are right. There's more legroom here than boasted by the Subaru, and easier heel and toe downchanges afforded by the perfectly placed pedals. The steering wheel, sadly and misguidedly smaller than those of earlier Integrales, still feels lovely to hold and look at, in a way the airbagged circular blob Subaru provides fails entirely to emulate. Climb aboard the Integrale, strap yourself in and you feel immediately ready for action. Do the same in the Subaru and you're as likely to feel inclined to head for the nearest Sainsburys.

Creating this homespun impression is the Impreza's best trick. Anyone could be fooled into thinking it a mildly warmed saloon, born to fight it out with tepid Vectras and Mondeos. And it is only when you have run it to the red-line in the early gears, looked in the mirror and seen just how far behind the Integrale lags that you realise that the Impreza, this Impreza at least, was born for rather greater things than it might at first suggest.

Make no mistake, a Subaru Impreza Turbo 2000 is formidably fast, proving an urge that's not only stronger than that of this fastest of road-going Integrales but one which arrives earlier and lasts longer, too. Side by side you'd never think the Subaru would be quicker but, in the event, it's not even a contest.

There are two factors at play ➛

Instruments galore and more legroom, but left-hand drive only for Lancia

Impreza cabin is clean and functional, but less than awe inspiring

here. Both cars deploy 2-litre, turbocharged, twin-cam 16-valve engines, the Lancia's noted most for its beautifully smooth manners thanks to its counter-rotating balancer shafts, the Subaru's for the rasping exhaust note resulting from its flat-four configuration. Of the two, it is the Integrale which is more powerful, providing 215bhp compared to the 208bhp of the Impreza; so you'll not find the source of the Subaru's extra urge here.

It's all in the kerb weight. Bald statistics show the Integrale bludgeoning the scales at a fraction less than 3000lb, while the Subaru, seemingly the bigger car, is fully 500lb lighter. Which means the more powerful Lancia has 159bhp for every tonne of weight while the Impreza provides 186bhp per tonne.

Which is not to say the Subaru has it all its own way. The Integrale engine, the final development of the twin-cam engine once found in the Fiat 132, is a fine motor, providing smoother delivery of its torque than the Impreza whose power can come in with a bit of bang if you are not careful with throttle. It also has fractionally the better gearchange though its innately tall and wide ratios suit the turbo engine less well than the broadly shorter-geared Subaru.

So far then, the Impreza has mounted a challenge against the supremacy of the Integrale as unlikely on the road as it has been on the special stage. Corners, however, present a rather different challenge.

The Lancia Integrale has always been famed for its attitude to corners and with good reason. This Evolution model, with its still wider track and stubby 205/45 ZR 16 tyres, propels the legend to rarely seen heights. What is remarkable is the way that it seems to shrug off its weight the moment a corner appears in its path.

Any race or rally driver will tell you that a heavy car is a cumbersome car, that weight, mathematically, militates against a car's ability to change direction and such weight slung out ahead of the front axle is the worst enemy of agility. The Integrale defies this logic. What is extraordinary is how easy it is to drive at supercar speeds with scarcely any effort. Hurtle up to a third gear corner, stab hard at the meatily reassuring brake pedal before turning in and the Lancia will simply flick on to your new course and stay there. The delay you expect from all such nose-heavy hatches as they swing into a corner seems scarcely to exist.

More than this, it will cling to dry bitumen to a point which seems scarcely believable for anything this side of a well-developed, mid-engined supercar and, in the wet, leave all two-wheel drive opposition for dead. Perhaps more importantly, the legendary Integrale steering remains as sharp as ever and its responses, once you finally find room to prod safely beyond the limit, are as predictable as you'd wish: it simply pushes its nose wide until you ease off the throttle, then snaps cleanly back into line.

The Impreza is truly humbled in the face of perhaps the most blinding blend of agility and stability yet to set foot on the public road. Drive it hard, use every morsel of its straight-line advantage and, true enough, the Subaru will stay with the Lancia on all bar the most twisting roads and, by conventional standards, you'd have to say it handled well.

Surprisingly, Italian left-hooker way out ahead for secondhand value

The standards here, however, are anything but conventional. Like the Integrale, the Impreza uses straightforward, strut-type suspension at each corner but it carries softer springing and a steering rack which, while using the same 2.8-turns across an undeniably tighter lock, feels decidedly less controlled than the super-precise Lancia steering.

As a result, the Impreza feels altogether more vague; a blunt instrument compared to Lancia's scalpel. Its smaller brakes may have less weight to control but, despite their undoubted powers of retardation, they never give quite the same air of breezy confidence boasted by the Integrale's fatter, wider discs.

Most of the time, it corners like a conventional hot hatch, albeit it one with fabulous traction. It heels into corners where the Integrale flicks, and feels restless in really fast curves where its rival seems impregnably clamped to the road, committed to following and relaying every change in the surface. Its 205/55 VR 15 tyres provide notably less grip than those of the Integrale and, once they start to slip, its reactions, while easy to contain, are never as clean as the Lancia's.

It does, however, have a trick up its sleeve. Find the space, far away from the public road, to really hurl it ➛

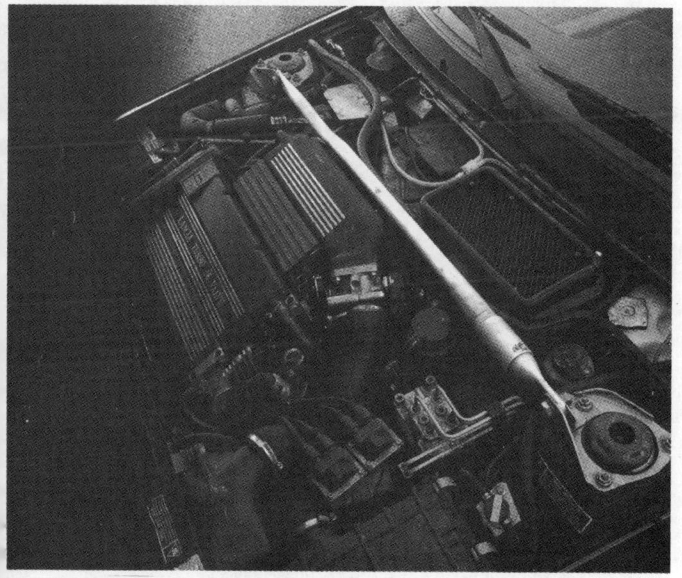

Classic Fiat twin-cam with 16 valves and turbo produces 215bhp

ubaru's 2-litre turbo develops 208bhp, but aluminium block saves weight

SPECIFICATIONS

	Lancia Integrale Evolution	Subaru Impreza Turbo
ENGINE		
Type	Four cyls in line	Four cyls, horizontally opposed
... ...	dohc, 4 valves per cylinder ...	dohc, 4 valves per cylinder
Construction	Iron block, alloy head	All aluminium
Bore/stroke ...	84 x 90mm, 1995cc	92 x 75mm, 1994cc
Fuel injection ...	Multi-point	Multi-point
Compression ratio	8.0:1	8.0:1
Max power ...	215bhp at 5750rpm	208bhp at 6000rpm
Max torque ...	220lb ft at 3500rpm	201lb ft at 4800rpm
Specific output	108bhp/litre	104bhp/litre
Power to weight	159bhp/tonne	186bhp/tonne
Transmission ...	Five-speed, manual	Five-speed, manual
Mph per 1000rpm	1st/6.2; 2nd/10.0; 3rd/14.2 ...	1st/4.8; 2nd/8.6; 3rd/12.2
	4th/18.8; 5th/22.7...	4th/17.2; 5th/22.5
CHASSIS		
Steering	Power assisted rack and pinion	Power assisted rack and pinion
Lock to lock ...	2.8	2.8
Brakes	271mm discs (front)	228mm discs (front)
... ...	242mm discs (rear)	230mm discs (rear)
Front suspension	Struts, lower wishbones ...	Struts, lower link
... ...	coils, anti-roll bar	coils, anti-roll bar
Rear suspension	Struts, transverse links... ...	Struts, trailing arms
... ...	coils, anti-roll bar	coils, anti-roll bar
Wheels	cast alloy, 7.5 x 16in	7 x 15in, cast alloy
Tyres	205/45 ZR 16	205/55 VR 15
DIMENSIONS		
Length	153.5in	170.1in
Width	69.7in	66.5in
Height	53.7in	55.1in
Wheelbase ...	97.6in	99.2in
Track (f/r) ...	55.0/55.0in	57.5in/57.1in
Kerb weight ...	2974lb	2463lb
PERFORMANCE		
0-60mph	6.0sec	5.8sec
Top Speed ...	134mph...	137mph
KEY ENGINEERING POINTS		
	Old Fiat engine kept on the pace with the aid of counter-rotating balancer shafts and a Garrett T3 turbocharger. The front to rear power split is governed by a torque-sensing (Torsen) centre differential. The first 4WD Deltas had 165bhp engines, rising to 185bhp for the first Integrale and 200bhp for the 16-valve version.	Now the legendary Alfa flat-four motor has been ditched, Subaru is alone in using this configuration today if you excuse the Mexican-built VW Beetle. Quad-cam motor is all aluminium helping towards-towards its lighter kerb-weight. Also departs from Lancia practice by using a viscous coupling for its centre differential, allowing an infinitely variable front to rear torque split
IN ONE SENTENCE		
	Every bit as good as it looks, in real terms one of the quickest road cars you can buy	Rather better than it looks; quicker than an Integrale but not so great to drive

round and suddenly it will start to behave like a not-so distant cousin of rally car. On a loose surface, it drifts beautifully through quick curves and, you time correctly the moment you nsettle the tail, it will slide into graceal oversteer into a hairpin and keep he back out of line all the way round, even when you come back ard onto the power.

Sadly, life on the open road is rarely unctuated by loose gravel surfaces nd, too often in our day with the two, was left longing to trade all the npreza's extra urge for so much as a ice of the Lancia's reactions.

Even so, by the end of the day, it was the Impreza which had improved most upon my expectations. What I had approached imagining to be not much more than a conventional but crude hot saloon, I left with new respect. Quite apart from its undoubted dynamic abilities, it is also a credible family car with a big boot and a reasonable amount of space inside. More importantly, I discovered a car which was genuinely enjoyable to drive hard for reasons other than the punch it packed between the corners.

An Integrale, however, it is not. While the Subaru feels like the ultimate performance derivative of a range of altogether more homespun machinery, which it is, the Lancia, despite relating to the Delta range in exactly the same way as the Subaru does to the Impreza range, does not; it feels like a purpose-built performance machine, one whose only purpose, it would seem, is to provide some kind of ultimate on-road thrill for the practically minded.

This perhaps explains one of the reasons why, three years from now, this Impreza, which costs £19,110 now, should resell for about £11,500 while this Integrale, which would have cost £23,249 three years ago, is now for sale for £21,750.

On the Portuguese Rally the Subaru Impreza challenged the legend of the Lancia Integrale and on the highways of Britain it has now done the same. The outcome in both cases was the same; the Subaru came close, closer perhaps than any would have betted but, in the end, it was the Lancia which emerged, still clutching its claim to be the finest road and rally car of our times.

Thanks to Richard Thorne Classic Cars, Reading (0118 983 1200), for the loan of the Lancia Integrale.

Subaru Impreza 2.0 Turbo 4WD 4dr v Volvo S40 T4

Volvo's family saloon fire-breather hunkers down to meet Subaru's all-wheel-drive rally winner

Volvo and Subaru both understand how well the gloss of motorsport triumph rubs off in the showroom.

The Swedes turned their image around with a string of touring car victories. Subaru's Impreza, even more impressively, won the world rally championship and acquired a cult status along the way.

All of which makes this battle between the new Volvo T4 and the Subaru Impreza Turbo an enticing prospect. Both are hot versions of more humdrum models and both use turbochargers to achieve power outputs that are neck-and-neck – the T4 delivers 200bhp, the Impreza, 208bhp. Prices are just as close, too, the Volvo is £20,125 and the Subaru, £19,715. But there is one important difference – like other Subarus, the Impreza has a full-time four-wheel-drive drivetrain, while the T4 relies on front drive tamed by traction control.

The T4 is the newcomer here, and gives the S/V40 range a hint of glamour – like the T5 does for its big brothers the S/V70.

For extrovert drivers at least, the Impreza wins in the showroom. Leagues more than the Volvo, it reeks of power and performance before it turns a wheel, flouting spoilers, oversize front fog lamps and a deep air intake nostril in the bonnet centre. Smaller spoilers and subtle alloys make the T4 far more discreet.

Mind you, the Impreza's looks are justified – it is quicker off the line, covering the 0-60mph sprint in 6.2sec, half a second faster than the Volvo. But the Volvo just shades it for top speed, 141mph plays 137mph.

Extracting the performance from each car is a vastly different experience. Volvo has tuned the T4's turbo for low-down response and refinement. It has 221lb ft of torque available from as low as 2400rpm, so you don't need to probe the upper reaches of the rev band, or drop down a gear, to swoop past slower traffic or to dial in more speed. Which is just as well because, at high engine speeds, the engine coarsens and our test car was plagued by a stiff, sticky gearchange.

The Subaru is more old-school turbo. Wind it up to 3000rpm and you'd think a switch has been flicked, such is the suddenness of your launch towards the horizon. Acceleration from 30-70mph is a fraction quicker in the Volvo, but the more free-revving Subaru streaks ahead at higher speeds. It feels twice as fast; a sensation aided by the inimitable growl of the flat-four engine.

But, much more important, the Impreza is the more satisfying to drive. Its four-wheel-drive chassis is memorably reassuring through bends, so you'll attack unfamiliar roads with verve. Even in the wet, the worst it will do is to run progressively wide of its line. If the steering was more precise, and had better weight and feel, you'd have little to wish for.

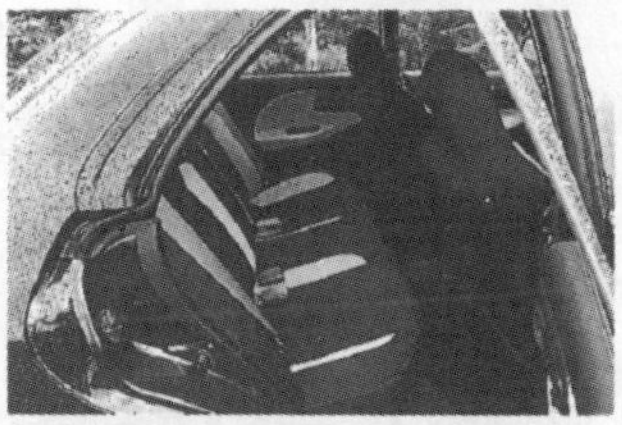

Cabin lifted by standard, two-colour trim; sports seat grips driver well

Roomier, more sober T4 has side airbags. Leather trim part of £1850 pack

Impreza dash looks dreary after exterior's scoops and spoilers. New steering wheel, white dials (not shown) enliven latest models

T4's cabin gives no hint of its performance potential. Wood-effect inserts make it far classier than workaday Subaru

The T4 isn't as intoxicating as the Impreza, but it's a fine drive nonetheless. It transmits its power through the front wheels far more cleanly than other turbocharged Volvos, and it clings much better through bends, too, ably assisted by an unobtrusive traction control system and good steering feel.

More often than not, the Volvo has the more compliant ride, too, especially over smaller bumps where the Impreza can crash down too harshly. But over faster crests and undulations the Suburu's superior body control is telling. It settles down more certainly than the T4 and is less flustered by a series of direction changes taken quickly.

Like the exterior, the cabin of the Impreza punches its sports credentials across hard. You snuggle into hugely bolstered seats, and, thanks to just-made improvements (old model pictured), you also grip a chunky Momo steering wheel and peer at white dials.

Sybarites will feel more at home in the Volvo, though. It's not just that it feels better put together than the Subaru, but its trim plastics are superior and the switches are better located, too.

Both cars are practical enough to satisfy as family cars. Either will accommodate four six-footers, although they'll have slightly more kneeroom in the rear of the Volvo. The Swede also has more luggage space.

Standard equipment is on a par: anti-lock brakes, central locking and electric windows all come as standard.

The Impreza offers twin airbags, while Volvo charges you £200 for a passenger bag. But then again, the T4 has standard side bags that are not available even an extra on the Impreza. Unusually, neither has a standard sunroof or air-conditioning, but both are cheaper options on the T4, which also adds a huge list of equipment packs and stand alone items the Impreza cannot match.

Running costs are close-run, too: the Impreza's insurance rating is group 17, the Volvo's a notch lower at 16. As for fuel economy, on the *What Car?* touring route the T4 returned 32.5mpg, edging ahead of the Impreza's 28.2mpg.

Lastly, the Volvo promises to depreciate slowly. But, there's a five-month wait for new Imprezas and secondhand ones sell fast, so it'll hold its value well, too.

Reason enough to buy. Just think though: fantastic power, apparently endless grip and sheer one-of-a-kindness. For us these are even better arguments for making the Impreza Turbo the one we'd lighten our bank balance for. ■

Make	SUBARU	VOLVO
Model	Impreza Turbo	S40 T4
VERDICT	●●●●	●●●
ESSENTIALS AT A GLANCE		
On-the-road price	£19,715	£20,125
Service cost for 3yr/36,000 miles	£1162	£1067
Ins gp/typ quote	17/£502	16/£408
Test mpg/Cost per mile	28.2/46.7	32.5/54.9
Max power bhp/rpm	208/5600	200/5500
Max torque lb ft/rpm	214/4000	221/2400
Max speed/0-60mph	137mph/6.2sec	141mph/6.7sec
30-70mph through gears	6.2sec	5.9sec
Braking distance 30/70mph	8.5m/52.5m	9.2m/49.9m
Space		
Front head/legroom	acceptable/good	acceptable/excellent
Rear head/legroom	acceptable/acc	acceptable/good
Boot capacity	11.0cu ft	16.9cu ft
Warranty details		
Anti-rust/paint warranty (yrs)	6/-	8/3
Warranty (months/miles)	36/60,000	36/60,000

The Mitsubishi Evo V and capable homologation cars

RAW like sushi

Subaru Impreza WRX are two of the tastiest, most breathtakingly we have ever driven. They are also two of Japan's best-kept secrets...

STORY BY **GAVIN GREEN**
PHOTOGRAPHY BY **ALEX PUCZYNIEC**

WILD LANCASHIRE MOORland, miles from anywhere, a dark stormy night. This is spooky country, just right for an alien sighting. And here they are: wings, huge mouths and banks of eyes radiating bright light. They're also moving very, very fast.

They're terrestrial aliens, of course, but they're so different from what we've seen so far in this country, Japan might as well be another planet. From Japan they come via a garage in Bolton. Warrender, which specialises in importing the weirdest, wackiest and least obtainable Japanese cars, has a forecourt stacked out in this sort of thing, winged, large-snouted and ostentatious. It specialises in Japanese machinery because, increasingly, the Japanese build the best high-performance cars. And because the official Japanese distributors in the UK seem pathologically incapable of selling the cars that enthusiasts actually want, there's an opportunity for a small, independent outfit to fill the void (and there are loads of other 'grey importers' apart from Warrender). Right now, rally look-alikes are what the smart young things in Japan want. And what the customer wants, Japanese manufacturers give – at least in Japan. Unsurprisingly, Mitsubishi and Subaru, the outstanding teams in international rallying today, dominate the market. And their latest offerings not only look like the sort of monsters that Colin McRae and Tommi Makkinen use to terrify small animals in the forests, they go almost as fast, too.

'The Mitsubishi has the most prominent chin we've seen since

Subaru, of course, serves up the normal Impreza 4x4 Turbo in Britain, an awesomely good car and probably the single greatest performance bargain we've got. But in Japan you can get it in WRX guise, which – astonishingly – manages to make the thing found in British dealers feel (and look) like a bit of a wimp. The WRX pumps out a claimed 280bhp. That's 72bhp above the normal Impreza Turbo, and 40bhp more than the excellent Prodrive version, modified by the people in Banbury who hand-make the world rally championship cars for Subaru. And anyway, 280bhp is the Japanese manufacturers' voluntary upper power limit: many reckon the WRX actually does better than that, on the sly. More powerful engine apart, the WRX has bigger brakes, a larger rear wing, and all sorts of extra kit to impress the gadget gurus in Tokyo, including a water spray button for the turbo's intercooler, essential when you're cruising Tokyo's Shibuya district of a Saturday night.

The other great mobile status symbol on the fashionable streets of Tokyo is the Mitsubishi Lancer Evolution Five (or Evo V for short). Again, it's a road-going version of the thing that rally stars use. Mitsubishi cars in the UK tend to be about as exciting as Mitsubishi TV sets and

Subaru: water spray pumps power to 280bhp

WRX looks like Colin McRae's factory wheels – nearly as fast, too

Once a very ordinary Lancer, now almost a rally car in drag

Mitsubishi: brand-name cabin and 8250rpm

Mitsubishi microwaves, which is why the fashion-conscious British youth of today are no more likely to aspire to own a Mitsubishi car than they would an M&S cardie.

Yet, in Japan, Mitsubishi sells the Evo V, a car with a competition pedigree, the look of a *Star Wars* interceptor craft, and enough fancy kit to distract Japanese youth from their Nintendo 64s. It, too, has an intercooler spray. It has vast Brembo brakes complete with red calipers and big 'Brembo' letters, a lovely Momo steering wheel, deep shoulder-embracing Recaro seats and big, open OZ 17-inch alloys ringed by low-profile Bridgestone Potenzas. All the brand names are highlighted. Even the Pioneer radio is called a Carrozzeria, the letters in bold gold. The Japanese, of course, love foreign brands, and the more Italian sounding, the better. The Evo V is like a mobile showcase for designer car brands, ideal transport for the Japanese twentysomething on the lookout for the next New Thing.

And they'll very nearly be buying the Real Thing, too, because the Impreza WRX and Evo V are quite similar to McRae's and Makkinen's forest wheels. For the Group N British national rally championship, simply fit a sumpguard and a rollover cage, and you're away. The works teams do a little more for the full-blown world rally championship cars (laptop-tuned engines offering more through-the-range power, water-cooled disc brakes, transmission 'brains' which bias torque split according to driver preference, seam-welded bodies for extra strength, etc etc). But the lineage is there.

Michael Schumacher strutted into F1'

They look amazing, especially the Evo V. It's plainly a very ordinary Lancer – similar to the European Carisma – under all the war paint and suit-of-armour aerodynamics, but the effect of all these add-ons is to transmogrify an ordinary little saloon into an extraordinary one. The nose is standard Lancer, down to the bumper. Underneath is very possibly the most prominent chin we've seen in motoring since Michael Schumacher strutted onto the GP scene. The mouth for the intercooler is a vast, square hole (the intercooler peeps through the mesh), and either side sit vast eyes, big flame-thrower spotlights, bright enough to turn dark moors into tunnels of daylight. Outboard of them, huge slats channel air into the front discs. From the front, the Lancer Evo V looks like some sort of vast insect-like monster, horribly aggressive.

The front wings are bespoke (from lightweight, high-tensile steel) and shroud the front Bridgestone Potenzas (225/45 ZR17s, the same as the rears). The bonnet has twin louvres, to aid breathing: they look like nostrils, only adding to the threatening, animate look. Along the sides are deep and angular skirts, accentuating the car's ground-hugging demeanour. The rear arches are flared, in a more tacked-on, cut-and-paste way. But none of those styling tweaks can hold a candle to that rear wing, a good foot-and-a-half proud of the boot, the best piece of aero kit we've seen since the original Escort Cossie. What's more, the alloy wing blade is four-place

'You have to have your wits about you. It may well be the fastest point-to-point form

Detailing on the Impreza verges on the dull compared with the wacky Evo. Ducts, slats, scoops and skirts abound

Rev-counter is mounted centrally

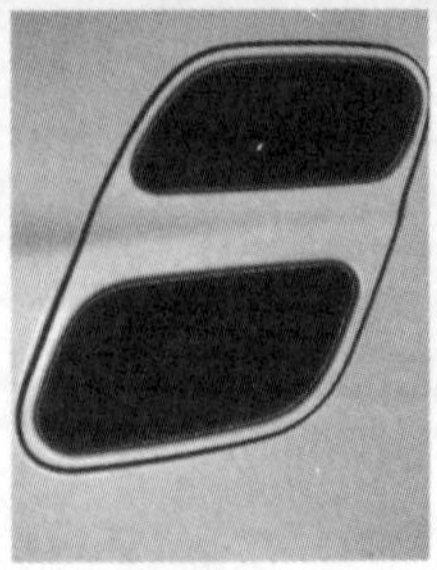

adjustable, so as to increase – or decrease – rear downforce as you motor through Tokyo.

The Subaru looks like a taxi alongside. It has trademark Impreza gold alloy wheels and a chin spoiler that's almost as deep as the Evo's, replete with big, square gob, hellraising spotlamps and slatted brake ducts. There's a bonnet scoop, too, just in front of the windscreen, to duct air into the intercooler, perched on top of the flat-four, red crackle-finished motor. And there are side skirts. You can see that it's not exactly a shrinking violet, then. But there are no flared wheel-arches, and nothing else dramatic about the flanks. Even the back wing is discreet (or at least it is compared with the Evo's). The 'WRX' green-and-grey typeface, plastered all over the rear end and doors, is also naff, more Fat Willy's Surf Shack than sports car.

Unlike the Evo V, there are no foreign brand names – apart from a Momo wheel – to increase the sushi bar cred. Everything, from the seats to the calipers (discreet black, rather than red) have plain 'Subaru' on them. This hurts the dynamics not one jot. In fact, the brakes and seats of the Subaru are a notch above the Evo V's.

In Bolton, people stop and stare and gawp. On to the moors, there's nothing but windswept hills and sheep and the best driving roads in the area. The Lancashire moors aren't as big or as empty as the North York moors or, better still, those in Durham, but they're woven over by challenging roads all the same. Our favourite, just south of Blackburn, is bumpy and winding, opening up into a smooth, fast-flowing road.

Both cars are immensely easy to drive. There is none of the Ferrari let's-start-again school of motoring, where every input is foreign to the driver of a normal hatch. They are upright, easy to see out of, easy gearshifting and tractable. Warrender's Steven Phillips didn't have to give us complicated driving instructions or a list of emergency breakdown numbers, just in case. The Evo V was brand, spanking new – even the seat covers were in place. 'Any running-in instructions?' I inquired. 'No,' answered Steven, with a 'why would there be any?' look on his face. There is often much to be said for the Japanese appliance school of car engineering. After all, what good is a car that has stopped working?

But you can tell, quite early in the relationship, that these are cars born of people with maths in their heads, not fire in their blood. The lovely cabin details, the deliciously communicative brakes, the sweet-talking steering, the progressive throttle action: they are the skills of men doing their job rather than engineers trained in art. There is little romance about these cars. But there is enormous speed and engineering integrity. In Bolton, everything is so easy; all that's weird is the vast tea tray in the rear-view mirror, and its delta base plate, redolent of a fighter jet's wing. But given freedom, the cars show their wild sides.

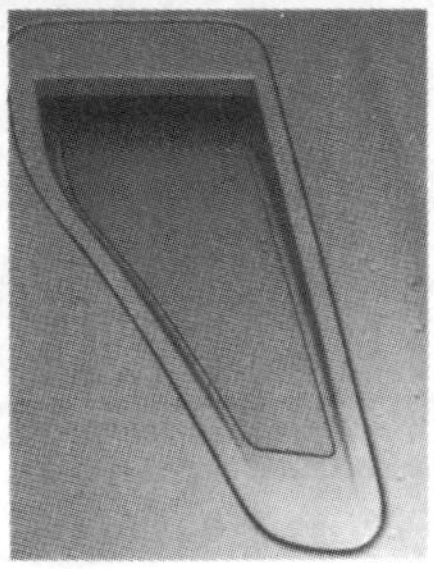

Evo dash is relatively uncluttered

No Evo panel has survived without being festooned in massive bulges and wings. Fortunately, its bite is as good as its bark

IMPREZA FIRST. ITS ENGINE SOUNDS rather quaint on idle, not least because the flat-four has a Beetle-like thrum. It does not suggest power. But tickle the throttle, and, between 3000 and 4000 revs on the white-faced tachometer, power arrives. It hits you like a wall of water. You don't so much drive to the horizon, as get hauled in by it. And while the car races forward, frenziedly, excitedly, the four-wheel drive and the big tyres and the brakes reassure you that you can continue to go fast, and faster.

The bumpy road jolts the car and makes for uncomfortable progress (in fact, the standard British-spec Impreza Turbo 4x4, and the Prodrive cars, have greater compliance). You have to have your wits about you, and it does end up being quite a wearying, physical experience taming an Impreza WRX here. But it will also leave you in absolute awe of what may well be the fastest point-to-point ground transport known to man.

The sensation, for the driver, is simply one of speed. You don't remember the man/car interaction, or the delicacy of anything. You remember brute power and an immense unfathomable ability to tame difficult terrain. And huge speed.

The engine is the one mechanical component with innate charm. It sounds good at low revs and it sounds even better as it bears down on its 8250rpm cut-off, howling like a mad thing, turbo spinning it to a frenzy. The little red-stitched black leather gearknob moves with a delightfully short, sharp throw.

The Evo awaits. After the Impreza, it turns out not to be quite as well-balanced when dancing over the moorland bumps and stringing together the bends. The bumps kick even harder, the suspension apparently even less yielding. The steering has more stiction and doesn't flow as sweetly, either. But its speed is just as great as the Impreza's. It storms over the road – traversing bumps, straightening corners, blasting over the tarmac. It isn't actually flying, but it's going so fast, and is so unconcerned by the terrain, that it might as well be.

Again, the pace is blistering, mystifying. A Ferrari or a Porsche would demand a relationship between you and the car to conquer this road. These cars simply require a guiding hand, and a modicum of competence. They represent science's conquest, rather than a driver's triumph.

Again, as with the Impreza, the Evo's turbo gathers boost from 3000rpm, which turns into a full-blooded wallop from 4000, singing on the turbo around to the cut-out 7500rpm. Barp barp barp as the limiter holds up its hand – you snick-snick it back a gear and the action starts all over again. The brakes are vastly powerful and, because they have ABS, they just stop, no drama.

It rains for part of the time we're playing, but the rain affects the cars not at all. Much the same speed, much the same high-speed indifference as in the dry. OK, OK, we can do this. Where's the next challenge?

HERE IT IS. WE HEAD SOUTH, NEAR Wigan, to the Three Sisters racing track – used mainly by karts but fine for rally cars used to tight conditions. Again, the Subaru feels better balanced, more communicative, easier to exploit. But the Mitsubishi is just as fast. They might both have 280bhp at the spec-sheet but, when they're really tanking on, they feel more powerful still.

It isn't so much the power that endears, though. It is the way they put it down on the road. Through tight direction changes, the cars are fabulously manoeuvrable (especially the Impreza), four tyres spreading all the power, four tyres helping to ease you out of mischief, superb brakes able to bring the whole show to a stop, if necessary, with amazing suddenness. You want to stop the game? Fine, just push the middle pedal. It's like a switch on a computer console. Game over.

Back in Bolton, Steven Phillips is telling us that he charges £31,000 for a WRX Type R Coupé STi (for Subaru Tecnica international) 'Version Four', as tested. It's the latest WRX, although there's already a new, hotter, 2.2-litre version out this month in Japan. Its full name is the Impreza Premium Sports Coupé 22B STi Version.

The Japanese have always loved prolix car names. Equally, they want the latest of anything, which is why cars (let alone computers and other electronic toys) are updated so quickly. Warrender's price for the latest Subaru WRX includes tax, MOT, and sorting out all the hoops that the personal import boys will make you go through. In Japan, it costs only £14,000.

At the moment, though, the latest Evolution V is *the* Japanese car to have in Tokyo. Mitsubishi originally said it would make 6000, when it launched the car in January, but has now bumped that up to 7500. Mitsubishi in the UK says it may import some, to sell through its dealer network, and is doing the sums (etc etc, excuses, excuses), but in the meantime Warrender will sell you one for £29,995 including all taxes, warranty, MoT and the rest of the on-the-road stuff. In

Japan, it's £15,000 – again, that's a staggering performance bargain.

Of the pair, I'd take the WRX. It has extra finesse, and also sounds sportier and more animate when the milestones start to fly by. It may be louder and more raucous – but who's going to buy one of these cars for long motorway drags? Besides, what do you expect when you're pulling 20mph per thou in top, all the better for serving up slingshot gear ratios?

The Evo V is the better pose, though. It looks more arresting and its kit is more fashionable. It's also the car the world rally champion drives. For young Japanese poseurs, gathering on a Saturday night, it has to be the car to buy. Equally, in Britain, the *Max Power* set would doubtless give a crateload of XXXX to have a go in one. It is the most dramatic saloon-based production car we've seen since the old Sierra Cossie, a car that looks almost as other-worldly as it goes.

Thanks to Three Sisters Circuit, 01942 270230

Specifications	Mitsubishi Evo V	Subaru Impreza WRX
Price	£29,995 (approx, UK)	£31,000 (approx, UK)
Engine	Transverse front-mounted dohc 16V four, turbocharged	Longitudinal front-mounted dohc 16V flat-four, turbocharged
Capacity (cc)	1997	1994
Bore/stroke, mm	85.0/88.0	92.0/75.0
Compression ratio	8.8 to one	8.0 to one
Power	280bhp at 6500rpm	280bhp at 6500rpm
Torque	274lb ft at 3000rpm	260lb ft at 4000rpm
Power-to-weight ratio	205bhp per tonne	224bhp per tonne
Transmission	Five-speed manual, four-wheel drive	Five-speed manual, four-wheel drive
Suspension, front	MacPherson struts, coil springs, lower wishbones, anti-roll bar	MacPherson struts, coil springs, lower wishbones, anti-roll bar
Suspension, rear springs,	Multi-link, coil springs,	Struts, lower wishbones, coil anti-roll bar anti-roll bar
Brakes front/rear	Vented discs, ABS	Vented discs, ABS
Tyres front/rear	225/45 ZR17	205/50 R16
Length/width/wheelbase	4350/1770/2510mm	4340/1690/2520mm
Weight	1360kg	1250kg
Maximum speed	112mph (limited)	112mph (limited)
0-60mph	5.1sec	4.8sec
Mpg	27.9 (Japan extra urban)	27.4 (Japan extra urban)

Classic Pick

Subaru Impreza WRX

Quinton Taylor sees a bright future for the Subaru WRX - the RS Escort of the '90s

The WRX RAs engine is a quad-cam, two-litre, 16-valve, horizontally opposed, fuel-injected, turbocharged and intercooled motor

Nowhere can you mistake the sound of it and local heroes Possum Bourne and Trevor Crowe are their most loyal supporters and developers. Possibly the car with the biggest following on the current rally scene, Subaru's Impreza took over where the Legacy left off.

The Impreza – with its reputation for reliability and sheer outright performance – is the RS Escort of the '90s.

It can be made to churn out around 600bhp in very special tune overseas but even local developers can get a reliable 400-450bhp out of the indestructible little boxer two-litre.

In the recent Cardrona Queenstown Goldrush hillclimb, Possum Bourne showed how quick and tough these cars are with a not-too-distant third in a car with only a third of the horsepower of the winners.

Subaru have even gone further with the recent release (albeit at a trickle) in New Zealand of the STi version, which is a real stormer.

The queue for one is very long!

Development

Subaru looked around for something lighter and nippier in the late eighties when the Legacy started to lose out to lighter and

Visually, they are a pleasant-looking little car, particularly with the standard deep spoilers and big front air inlets coupled with a medium-sized boot spoiler

smaller cars such as the Mitsubishi Evo Lancer series and Ford Cosworth Escorts and the result was the Impreza that has done phenomenally well in world rally circles.

The engine is now a four-cammer, two-litre, 16-valve, horizontally opposed, fuel-injected, turbocharged and intercooled motor – the result of a continuous development programme to overcome problems of cracked pistons and breakages that used to sometimes occur when the old Legacy Rally cars were pushed to their limits.

Now it's rare when something breaks on these tough little cars and it's usually only when some roadside obstacle has been clouted that they stop.

Ventilated disc brakes are fitted all round if you are lucky enough to score an 'RA' version, while gas struts and stiffer suspension bushes all make for a rather firm ride.

The interior is nothing spectacular and is all sombre black with a Momo wheel fitted as standard. Even the control panel and dash has little to excite you visually. It's just all function and long-lived like the rest of the car.

Visually, they are a pleasant-looking little car, particularly with the standard deep spoilers and big front air inlets coupled with a medium-sized boot spoiler and polished alloy wheels shod with fat Michelins.

The body colours are basic, too, with little adornment other than small areas of black paint and a couple of huge driving lights mounted low in the front air-dam. Companies like Crowesport market a wide range of hot-up and dress-up parts for these cars to make them very distinctive according to personal taste.

Driving a future classic

I recently had the pleasure of being lent a WRX to sample, an early model imported from Japan, on the market at about $25,000. It had obviously done a bit of hard work but was generally in pretty good condition, and at the price was good value for a very formidable road car.

I'm picking that this model will be a classic and if you can get a good one that hasn't been too thrashed they are an excellent buy. Particularly check if the rev limiter has been played with as the pistons tend to develop little rattles if 8000+ rpm is used too often!

Subaru have always had a reputation for top build quality and their engines are usually built to something close to aircraft tolerances. My brother-in-law is employed by a local agent and delights in working on them as their construction is straightforward and is always top-quality engineering.

Fortunately for enthusiasts everywhere there is such a thing as homologation, which resulted in a road version of the Impreza rally car, the WRX, being marketed and the initials themselves have become immediately associated with the name Subaru.

A WRX, in any shape or form, is a very forgiving car to drive. Modified as they are, they certainly offer somewhat harsher riding but not uncomfortably so. The sample driven had 30 something kms on it but was relatively free of rattles and still felt very taut.

The controls are straightforward and there is nothing really that could be described as fiddly. Hit the accelerator and pick some good roads and the Impreza is a real delight with some 180kW of power in a light car plus four-wheel-drive making for some impressive cornering power. Subaru have set the WRX up to handle pretty neutrally and this it does with a combination of superb steering and ABS brakes really putting the driving pleasure back into driving.

The WRX still has that Subaru bark, albeit rather subdued for road use and without the crackle of the wild computer management system of the rally cars.

Acceleration in the dry is phenomenal with a 0-100kph time of around six seconds depending on the state of tune.

In the wet, full-bore acceleration really has the diffs working to keep the handling under control but it seldom gets out of shape unless you are doing something really stupid. Cornering is absolutely flat and controlled, making a mockery of road speed advisory signs for corners!

Looking ahead

Put one of these away now and use it for those weekend trips up country. A new one will set you back $45,000 plus, but there are bargains around; however, get them checked by an acknowledged expert.

Routine maintenance isn't too bad with everything easily accessed, however, you do need the assistance of somebody well versed in electronic wizardry as there is quite a bit of it in the WRX.

Comparatively economical under normal use, they will devour petrol shamelessly when wound out but then you are getting V8 performance from this little boxer. Handling is an absolute delight and they are built as tough as anything around and a lot better than most.

Parts are no problem from Subaru as even my old chugger coupe has a full parts list from Subaru, evidence of the recognised owner loyalty of this company.

They are one of the few cars out of Japan that have character and will in time deserve to be rated among other classics.

QUINTON TAYLOR

Subaru IMPREZA

MODEL TESTED WR Prodrive **PRICE** £30,169
TOP SPEED 141mph **0-60MPH** 5.6sec
30-70MPH 5.7sec **60-0MPH** 2.9sec **MPG** 21.2
FOR Sweet handling, blindingly quick across country, exclusivity
AGAINST No quicker than standard car, high price, weak brakes

STAN PAPIOR

New ECU, Ramair filter and sports exhaust bump 2.0 flat four to 240bhp

Prodrive-fettled Subaru Imprezas are easy to forget about among the tide of grey import WRX models sweeping on to our shores. But in the meantime the works Subaru rally team is still winning rallies and producing its own range of outstanding road cars based on the standard Impreza Turbo. In all honesty, a Prodrive Impreza like this WR model is the only option for those people who consider a WRX or Mitsubishi Lancer Evo V too much of a risk. Vitally, the modifications don't invalidate your warranty.

Quickshift is, er, quick but heavy

Links with world rallying are real

There are four elements to the Prodrive treatment: a body kit (£2467), a chassis tweak (£3231), an interior rethink (£2820) and an engine tune-up which raises power to 240bhp (£1968). Our test car had the lot, upping its price by a wincing £10,163 to £30,169.

One of the greatest strengths of the standard Impreza Turbo is its value for money. Which means Prodrive's interpretation needs to be a revelation to justify the extra outlay.

An obvious place to begin the debate is in a straight line. The now familiar 1994cc flat four engine is mildly tuned with a revised ECU, Ramair air filter and a new exhaust. Prodrive claims power leaps to 240bhp from 208bhp, with a similar increase in torque (now 240lb ft at 4000rpm). Our test car was also fitted with a quickshift (£260), which shortens and tightens the gear lever throws. It's almost too heavy now.

A blip on the throttle reveals an engine note that's louder and throatier yet more cultured than usual. Your first drive down the road will have you reaching for the window switch to hear more.

Extra row of dials in centre of dash

It may sound wonderful, but ultimately the WR proved no quicker than the standard Impreza Turbo.

Make no mistake, the Prodrive machine is plenty fast.

Standard car's air intake still there

Extra support from Recaro seats

WR overflowing with adjustability and feel – just like standard car

Any saloon that is capable of reaching 60mph from a standstill in 5.6sec and 100mph in 15.9sec is not disappointing. But the fact is, the standard five-door Impreza reached both increments 0.1sec faster when we tested it in February. The result is the same between 30 and 70mph (WR 5.7sec, regular hatch 5.6sec). And the Prodrive car is 2mph slower flat out (141mph against 143mph).

Weight isn't an issue, either; the WR saloon weighs 1283kg to the standard five-door's 1306kg.

Prodrive claims to have uprated the brake pads to reduce fade and shorten stopping distances. They certainly offer terrific pedal feel, but their performance is actually slightly inferior to that of the standard car.

The saloon's weight advantage over the five-door may also account for its slightly better fuel economy figures. The WR averaged 21.1mpg (against the hatch's 19.5) and returned 30.7mpg over our touring route (30.2).

Fitted with uprated springs and dampers and 17in Speedline wheels, the WR is an inspiration over a challenging road, with levels of composure, adjustability and feel that few cars can match. Our test car had been set up for ultimate stability, resulting in more understeer than normal, but Prodrive can tailor the chassis to suit. And the ride was less harsh than the regular car's.

This is a great car, no question. The trouble is that it doesn't make enough progress over the standard Impreza Turbo to justify the 30 per cent hike in price – objectively, at least. But the cachet of driving a car tailored to your requirements by Prodrive will undoubtedly be worth an extra £10 grand to some. And it's unlikely that anyone else could do a better job.

If you want power oversteer, Prodrive can set up the chassis to oblige

1 Standard car's leather-wrapped Momo wheel is wisely retained 2 White-faced dials standard, but WR gets extra ones 3 Noise well suppressed despite frameless windows... 4 ...which are powered, of course 5 Standard seats replaced by grippy Prodrive Recaro Sport items

High rear wing for WR saloon

PERFORMANCE AND SPECIFICATIONS

ENGINE

Layout 4cyls horizontally opposed
Max power 240bhp/5600rpm
Max torque 240lb ft/2400rpm
Specific output 120bhp/litre
Power to weight 187bhp/tonne
Torque to weight 187lb ft/tonne
Installation Front, transverse four-wheel drive
Construction Alloy head and block
Bore/stroke 92.0/75.0mm
Valve gear 4 per cyl, dohc
Compression ratio 9.5:1
Ignition and fuel Hitachi ignition, sequential multi-point sequential fuel injection

GEARBOX

Type 5-speed manual by Subaru
Ratios/mph per 1000rpm
1st 3.17/5.1 **2nd** 1.89/8.6
3rd 1.30/12.4 **4th** 0.97/16.6
5th 0.74/21.8
Final drive 4.44:1

SUSPENSION

Front MacPherson struts, lower wishbones, coil springs over dampers, anti-roll bar
Rear MacPherson struts, coil springs over dampers, anti-roll bar

BRAKES

Front 280mm ventilated discs
Rear 245mm ventilated discs
Anti-lock Standard

STEERING

Type Rack and pinion, power assisted
Turns lock to lock 2.7

WHEELS AND TYRES

Wheel size 7Jx17in, cast alloy
Tyres 205/45 VR17 Pirelli P-Zero
Spare Full size

WEIGHT

Kerb weight 1283kg
Distribution f/r 59/41 per cent
Max payload 530kg
Max towing weight n/a

MAXIMUM SPEEDS

5th gear 141mph/6470rpm
4th 116/7000 **3rd** 87/7000
2nd 60/7000 **1st** 36/7000

ACCELERATION FROM REST

True mph	sec	speedo mph
30	2.0	31
40	3.0	41
50	4.2	51
60	5.6	62
70	7.7	72
80	9.8	82
90	12.3	92
100	13.9	102

ACCELERATION IN GEAR

mph	5th	4th	3rd	2nd
10-30	-	-	6.9	4.2
20-40	15.4	8.6	5.2	3.0
30-50	13.2	7.3	4.0	2.4
40-60	11.2	5.7	3.5	2.6
50-70	9.1	5.0	3.4	-
60-80	8.1	5.2	3.7	-
70-90	8.0	5.4	4.6	-
80-100	8.5	6.0	-	-

Standing qtr 14.4/96mph
Standing kilo 26.1sec/119mph
30-70mph 5.7sec

BRAKING

From 30/50/70/96mph 10.0/25.8/50.4/95.7m
60-0mph 2.9sec

NOISE

At idle/30/50/70mph/ max revs in third 56/71/72/75/82dbA

FUEL CONSUMPTION

Government figures
Urban/extra urban/ combined 20.5/34.9/29.7
Autocar figures
Test/touring mpg 21.2/30.7
Best/worst 9.2/30.7
Tank capacity 60 litres (13.2 gallons)
Range 405 miles

Performance figures were taken at the Millbrook Proving Ground with the odometer reading 14,200 miles. AUTOCAR test results are protected by world copyright and may not be reproduced without the editor's written permission.

LAYOUT

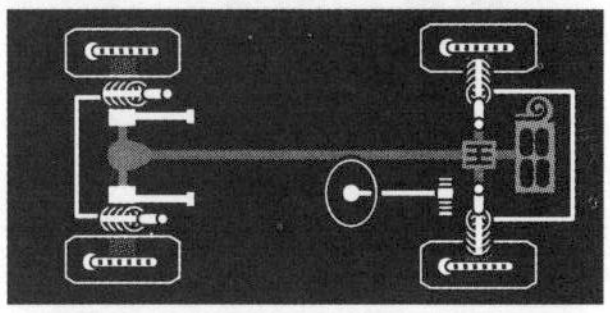

WHAT IT COSTS

On-the-road price	£30,169
Total as tested	£30,169
Cost per mile	n/a

EQUIPMENT

(**bold type** denotes options fitted to test car)

Automatic gearbox	-
Driver/passenger	●/●
Anti-lock brakes	●
Metallic paint	£241
RDS radio/cassette/CD changer	●/●/£298
Air conditioning	£1550
Electric sunroof	£799
Front/rear electric windows	●/●
Alarm/immobiliser	●/●
17in alloy wheels	●

● standard - not available

INSURANCE

Group	18 (est)
Typical quote	£620 (est)

WARRANTY

3 yrs/60,000 miles, 6 yrs anti-corrosion

SERVICING

Major 15,000 miles, 2.0 hours, £128 parts
Interim 7500 miles, 0.9 hours, £72 parts

Body 4dr saloon **Cd** 0.35 **Front/rear tracks** 1465/1450mm **Turning circle** 10.4m **Min/max front leg room** 910/1110mm **Min/max rear leg room** 630/840mm **Max front head room** 980mm **Max rear head room** 950mm **Interior width front/rear** 1410/1410mm **VDA boot volume** 365 litres/dm3 **Boot length** 960mm **Boot height** 380mm **Min/max boot width** 960/1330mm

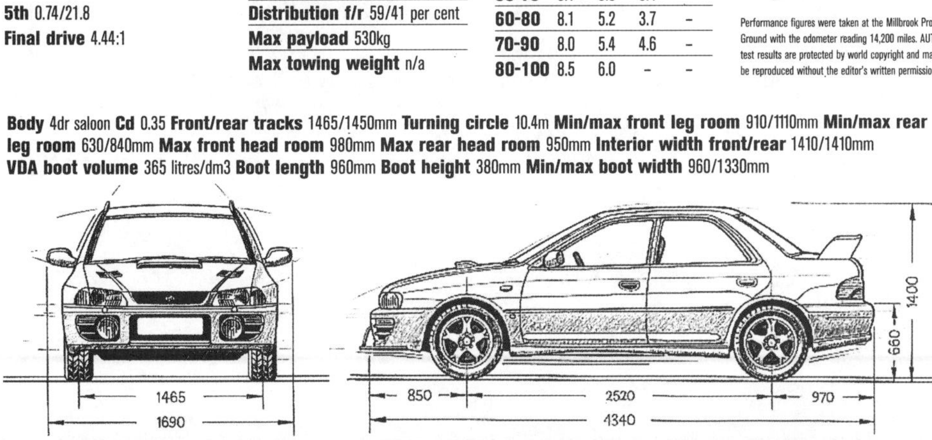

Sensational, but standard car is better

Wide Ra

ly Clone

Until now the closest any of us have got to being Colin McRae is on a Playstaion. Now thanks to the Subaru Impreza STi 22B we can experience WRC thrills on (or off) the nearest highway. CCC goes ballistic in the looniest Impreza yet

Left: forget the virtual reality world of the PlayStation because now you can have the real thing in real time. Just the the thing for when you come over all Colin McRae

The Subaru Impreza Turbo is the car of the late '90s. Now, in through the back door, the ultimate has arrived. It's the street version of Col's rally car - the Impreza STi 22B – and lordy is it quick

You've seen the video, you've read CCC, you've played the PlayStation game, you may even have bought the T-shirt – now Subaru is offering you the ultimate virtual reality Colin McRae experience – the Impreza STi 22B.

You'll have seen Impreza specials before, of course – there seems to be a new one released every other week. But this one is special, very special. Because this one is a replica of the Peter Stevens-designed WRC car in which Our Col has been doing kinda well in this season. It's not an exact copy, granted, but it's good enough to stun your mates down at the local car club and is guaranteed to empty out the bar if you park it outside your local boozer.

To be produced in a limited run of just 400 (all of which have already been sold, goddammit), the 22B has a two-door bodyshell, full works arches and front and rear bibs, and that monster tail spoiler. All that's missing on the outside are the 555 stickers and a bit of gravel rash.

Closer inspection reveals a few minor differences which prove you haven't picked up the four wheel-drive super-Scoob straight from the Prodrive workshop. The wheel arches are a slightly different profile to the WRC car, so that the 17in BBS road wheels stand a chance of filling the arches. The wing on the boot is taller

The quickest thing on the road? Quite probably. Big arches are slightly reprofiled to accomodate 17in BBS rims. Rear wing taller for visibility

than Colin's so that you can see what's happening behind you (obviously not a concern for Mr McRae), and the front and rear aprons are similar in style, but not identical. The bonnet vents are modelled on those of the WRC, but they're smaller, in order that they fit the standard bonnet cut-outs. Oh, and when you tap the panels you'll find they're steel, rather than carbon fibre – well, Subaru had to save money somewhere.

Subaru's STi (Subaru Tecnica International) outfit in Japan is responsible for producing the 22B, but Prodrive (though no-one there will admit it) developed the suspension set-up – hence the car's Bilstein dampers and Eibach springs. Each 22B rolls off the production line as a 2-door WRX coupé, before STi gets hacking and fettling and turns it into a 'modified' motor. The car's registration document even says it's modified.

The pumped-up arches add an extra 80mm to the Impreza's width, which in turn allows the front track to grow by 20mm and the rear by 40mm. A little clue that Subaru's rally maestros may have had a hand in the suspension development is the fact that the specially-made Bilsteins actually wear Prodrive stickers. All the suspension bushes are upgraded and the lateral suspension links at both ends of the car are rose-jointed. And there's a quicker-ratio rack for the power steering.

Rumours persist, though strenuously denied by Subaru, that the 22B has 350PS, despite its quoted figure being 280PS. But reliable sources reckon that the claimed output is the pukka one, which is a bit disappointing given that other STi Scoobies have the same peak power.

Above: 280PS or 350PS? No one's quite sure! Capacity is up, though, from 2-litre to 2.2-litre with handy rise in torque

That said, STi has put some hours into the Impreza's flat-four motor to give it more wallop than its siblings can muster. Capacity rises from 1994cc to 2212cc thanks to a bigger bore, and while the power doesn't rise, the torque output does. It's a seemingly modest increase, from

Right: Intercooler water spray button. Below: All you need is the decal pack!

260lb ft to 267, but the peak is now at 3200rpm rather than 4000, and most of it's available (in the form of thumping acceleration) between 2800rpm and 5200rpm.

Although little more than a gimmick in a road car, the 22B has water spray cooling for its intercooler to help keep intake air temperatures lower during extended hard use – but can anyone tell me where you can use a road car that hard for that long on Britain's public highways?

Public highways are where you'll enjoy the 22B though. It's a shame there's no roll cage to climb through on your way to the driver's seat, but doubtless some folk will fit one. Apart from a limited edition plaque on the centre console, the 22B's interior is much the same as any other STi's. Except that you can see those fat arches poking out in front and behind.

With the familiar throb of the Subaru flat-four pulsing past your eardrums, you snick the short-shift lever up into first and let out the clutch. It's a special twin-plate item (one of the plates is ceramic), and unless you're very positive letting it back in again, you can feel the plates juddering as they start to mesh. As the torque feeds through the transmission and heads earthwards, you can feel and hear the 235/40 profile Pirelli P Zeros chewing into the Tarmac, and you know immediately that finding traction won't be a problem.

Though violently quick from standstill – it'll hit 60mph in less than 4.3secs, 100mph in less than 11.9secs – the absence of drama can initially fool you into thinking that the 22B is slightly ordinary. Wrong! Introduce it to your favourite stretch of snaking back-road and you'll get the full benefit of the car's massive poke.

No matter how short the straight, by the time you hit the brakes hard for the next corner, you'll have punched your way up to at least fourth. The 22B's got very short ratios anyway, but thrust arrives with such speedy brutality that you're swapping cogs as fast as on a motorbike; it's like firing off a pump-action shotgun when you're trying to do away with a Terminator.

The 22B's torque is as impressive as its gunshot power. Fifth gear lets you accelerate as briskly as third gear in a hot hatch. Getting caught in the wrong gear is seldom a problem and there's the additional bonus of no turbo lag.

Fortunately, the brakes are as gutsy as the motor, allowing you to brake insanely late into a corner. And talking of corners, you can forget trying to flick the 22B sideways, on tarmac at least. This car's got phenomenal grip – it's a challenge just to get the tyres to chirp, let alone slide around at either end. Maybe it's possible to encourage some lateral action by playing with the driver-controlled torque split, but as our 22B was about to be delivered to its new owner, such experimentation wasn't really possible.

Prices for the 22B vary from between £40-50,000, depending on which importer you speak to, but as all the cars are already sold, expect values to now climb sharply.

It's a helluva car, the 22B; there are few new cars which will get you from point A to point B any faster, regardless of class or pedigree. Acceleration junkies will love it to pieces, and owners of supercars will give it a wide berth for fear of humiliation.

SUBARU IMPREZA WRX STI 22B

Price: £40,000 or more

Engine: 2212cc all-alloy 16v turbocharged and intercooled flat-four, 280PS at 6000rpm, 267lb ft at 3200rpm

Performance: 112mph artificially limited, sub-4.3sec 0-60mph, 20.1mpg

Wheels and tyres: BBS 8.5J x 17, Pirelli P Zero 235/40ZR17

THE 1999 SUBARU IMPREZA TURBO WON THIS comparison at 11.32am on Friday morning, just outside Reims on an arrow-straight French autoroute. There wasn't any dissent in the car; it was all very amicable and grown-up. We all agreed – me, photographer Gus Gregory and (*CAR*) designer Farhan Baig – that the Impreza was clearly better than the Audi TT, probably better than anything we'd ever driven in our lives, and that was that, nothing more to be said.

Now all we had to do was actually drive the Audi. Such is the enigma of the Impreza Turbo that it beats cars you haven't even driven yet. Three-up at 120mph, with the husky flat-four turbo drilling through solid kilometres to Munich where the TT was waiting, the magnitude of the Subaru's brilliance was overwhelming. It had been a few months since I'd driven one, and I was enraptured anew by that engine, the speed, the balance, the steering. No wonder this car is considered one of the greatest of the '90s, no wonder it's had so much praise heaped upon it. No wonder it is now much more than a Japanese saloon car: it's a cult.

By which I mean it has a special following; when you say 'Impreza' you're talking about more than mere metal. You're talking about the centrepiece to a whole new car culture, where Japanese tech, computer games, world rally championships and a

fashion for private imports all meet on 16-inch wheels. But cults are closely linked to fashion, and fashions tend to go out of fashion. So what's next?

Subaru has developed a new '99 model, with more power (up from 208 to 215bhp), new wheels, spoilers and interior trim to keep its car competitive, but it can't stay top dog forever, surely. Despite our very professional and thoroughly researched verdict on that long drive to Germany, the Subaru's place on the map must be invaded eventually. And if there's one car out there on the horizon that might challenge it, it's the new Audi TT.

Here's some reasoning: the TT has a turbo four-cylinder (like the Subaru), four-wheel drive (like the Subaru) and 225bhp (which is a lot). Also, it has a kind of *masculinity* about it – the ugly phizog, the tank-like slab sides. It's not a beautiful car as such, but beautiful cars can't become cult cars. Don't ask me why. I just decided.

I've also decided that slow cars, cumbersome cars, so-safe-to-drive-it's-impossible-to-crash-them cars, they can't be cult either. You need a bit of rawness, a bit of danger – and that's why we're spending a winter weekend in the bitter cold concrete of Munich (-9degC) to find out. Is the TT enough of a driver's car to replace the Subaru? Or is it all Bauhaus and no beef?

WE MEET THE TT IN AN UNDERGROUND CAR PARK, where it looks absolutely at home. It is small, really small when you stand next to it, a chunky little nugget of a car, with shallow side windows and a low, arching roof line. Suddenly the upright Subaru with its big glass area looks a bit dumpy.

It also looks like a concept car, someone's idea of how cars should look in a 21st century machine age; a kind of re-found Modernism amongst all the post-modern retro stuff. It's a car you can just look at. And look at.

But, hey, why look when you've got the keys in your hand? Ten minutes later we're doing 120mph on an autobahn.

First important point. You may have read that the TT isn't that quick. I know I had before travelling to Munich, and I think I know how people came to that conclusion: this 1.8 turbo has a torque curve that's so flat (206lb ft from 2200 to 5500rpm) that the acceleration is long and sustained, with no jolt or kick or sudden surge, like the Impreza's lag-lag-ignite take-off procedure (igniting 214lb ft at 4000rpm, incidentally). The TT just goes and goes, but it disguises its going in a smooth, steady delivery.

But that doesn't mean it's not yee-ha hold-on-tight overtake-

AUDI TT

£30,000 approx

1781cc 20V dohc turbo four

225bhp at 5900rpm
206lb ft at 2200-5500rpm

Six-speed manual, four-wheel drive

MacPherson struts, lower wishbones, anti-roll bar

Combined trailing and double lateral control arms, anti-roll bar

1285kg

150mph

6.3sec

everything fast. In fact, exhaustive research on the autobahn (it's a job) proved a remarkable thing: the TT is actually faster than a '99 Impreza. First of all in second gear on a slip road, and then in fourth gear (fourth out of five in the Subaru, one of six gears in the TT) in lanes two and three of the autobahn, we drove side by side and then nailed it. The Audi slowly but clearly pulled a bonnet ahead, and on one open stretch, the pedal to the floor in sixth pulled us all the way to 230km/h in one, hard-nosed, no-nonsense surge. That's an indicated 140mph, folks. So let's drop the 'TT not fast enough stuff, shall we?

Considering the TT is really a Golf, its components have got an awful lot of character. Character is the Subaru's ace card – the ex-traordinary flat-four that's like nothing else on the road, the promi-nent turbo whistle, the deep, race-derived seats that grip you from your thighs all the way to your shoulders, the Momo wheel: like a Volvo V70, you could drive an Impreza blindfold (not recom-mended) and recognise it.

The danger of the TT is that you can't distinguish it from an-other Golfseatskodaaudi platform-shared mush. But the TT is clever car. Have I mentioned the interior yet? How a mass manu-facturer manages to build cars of this quality and with these detail is beyond me (and most other European manufacturers) – forge Mercedes-Benz, this is the new quality benchmark. And everythin

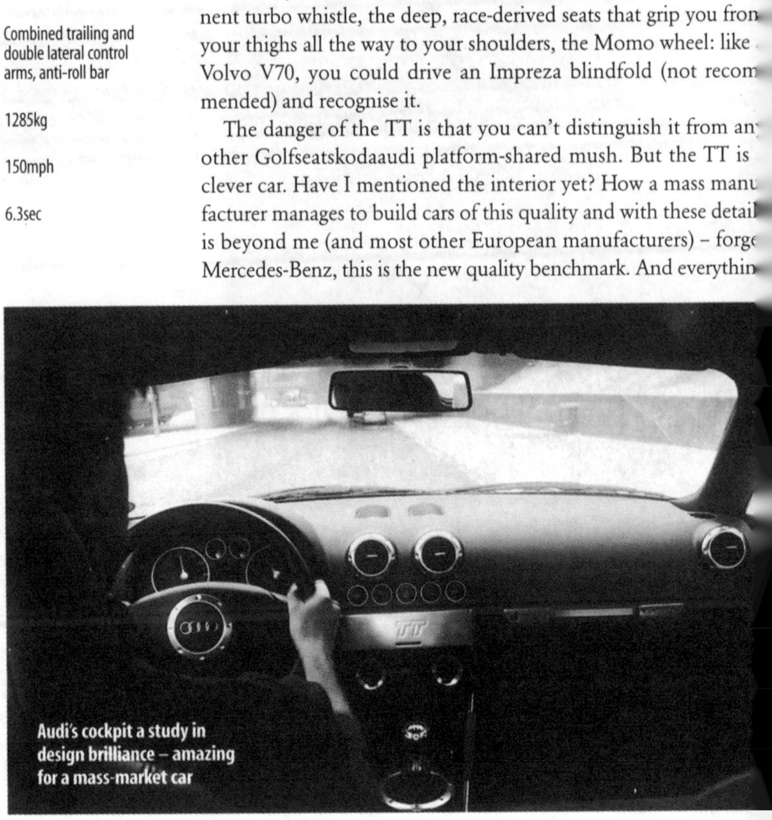

Audi's cockpit a study in design brilliance – amazing for a mass-market car

you touch feels right, from the excellent three-spoke steering wheel, to the vents, to the rubber gate around the gearlever. Even that engine, a specially blown Golf GTi unit, sounds throaty under power, surprisingly boomy on the over-run. It never feels or sounds mundane. The seats don't grip as well as the Impreza's, and the visibility is poor compared with the Subaru's, but this is the interior you'd want to sit in, rather than the plasticky Japanese effort. The Subaru's new white-faced dials (Japan-spec WRX-stylee) redeem it a little, but generally speaking it's cheap to look at, cheap to touch.

That Audi steering wheel is important. The Subaru's Momo wheel is the gateway to a wonderfully exploitable chassis. The steering is pointy without being nervous, offering total 4wd stability in perfect harmony with throttle-sensitive tail-slides through empty roundabouts (instructions – just go in fast and back off sharp).

And it's all down to two overwhelming impressions that anyone, from your kid brother to your old gran, could recognise – lightness and balance. Neither the front nor the rear of this car are in control; it moves all of a piece, all four tyres gripping or slipping according to your strict instructions. Eager, light-footed and fleet, it inspires utter confidence.

This is where the TT shows its more humble mechanical origins. Its steering is probably the best ever in an Audi, but its weight, though welcome compared with over-assisted woolliness, is more

SUBARU IMPREZA TURBO

£20,950

1994cc 16V turbo flat-four, dohc per bank

215bhp at 5600rpm
214lb ft at 4000rpm

Five-speed manual, four-wheel drive

MacPherson struts, coil springs, transverse links, anti-roll bar

Independent struts, trailing arms, dual links, coil springs, anti-roll bar

1235kg

143mph

6.3sec

Subaru's dull plastics let you concentrate on that white-faced speedo

artificial than the Subaru's, as though it's a product of a power-steering tweak rather than real road resistance. The TT also feels heavier at the front end, less nimble to turn in than the Impreza. That doesn't mean it's cumbersome or wallowy, but it lacks the instant precision of the Subaru.

What it (happily) doesn't lack is playfulness. That back-off-mid-corner test is a bit hackneyed now, but with a car like this it's important – it proves the car isn't inert, safe to the point of boredom. The TT will go sideways, just like the Impreza, when you ask it (then, and only then). The rear end breaks away so gradually, and the steering is so honest (and just 2.5 turns lock to lock) that it is, like the Subaru, great fun – never alarming and very predictable. Which means not only can you drive really fast in the TT, you can also challenge yourself on the way. Whoa, hold up those nine-point-nines – big score to Audi there.

So, the Audi TT is worthy of all the attention it's getting, and that ripple of excitement that's saying, 'Look, here's a new player, a serious contender.' At least, it *will* be a contender when it arrives here in late '99. But does it knock the Impreza off the top cult car spot?

I'm grimacing now, and staring at the ceiling, drumming my fingers on the desk. Hell, I've been thinking about this moment for days, and here it is – the decision. Which is better?

I think it's the Subaru, still. There's a sense of knowing behind the Impreza. It's not obvious why you should like it – it's kind of ugly, and ordinary in a way, despite the new wheels and new rear wing. But you feel like you're 'in the know', and of indisputably good automotive taste when you drive one. The Audi, on the other hand, is the result of millions of pounds spent on handcrafting a design that will be liked and admired by you and your friends. Just look at it – it's not as contrived as, say, the Fiat Coupé, but it still turns heads everywhere, and it's still a very deliberate bit of styling. The rally-derived Impreza is accidental, more honest – its bonnet vents are *needed*. Aluminium interior heating vents are not. It doesn't have an aluminium flip-up fuel filler surrounded by allen bolts. It does have very big driving lights.

Alongside the TT, the Subaru looks so unfashionable that maybe it won't go out of fashion after all.

But I'm torn in saying that, because I love the TT a lot, and I loved driving it. Look at it this way: being close to an Impreza means being close to a car enthusiast's idea of perfection, an automotive immortal. The Impreza is that high a benchmark. And the TT is close. Really really close.

SUBARU IMPREZA

[1994-present]

Subaru has years of experience of making high-performance four-wheel-drives that look like family shoppers but go like desert racers, and the Impreza Turbo is the best of the lot. Its conventional styling fails to disguise the rally breeding that makes it so exciting to drive. Despite being happy to bumble around town, the Impreza can hit 60mph in six seconds, with a 140mph top speed. But that's only half the story. Superb grip and handling make it unbelievably fast across country in expert hands. Drive an Impreza Turbo and its all-round excellence puts the performance of many more expensive cars into a new, unflattering light. High demand and short supply ensure that Subaru prices are as solid as the construction, and running costs are sensible. For all-round real world ability, there really is no contest.

Richard Wilsher

STAN PAPIOR

VITAL STATISTICS

PRICE RANGE £12,000 (private 95M) to £22,500 (private 98S)
LAUNCH DATES Flat-four and five-door, 211bhp, March 1994; facelift, October '96; more low-down torque, new facia with twin airbags, October '97; 218bhp, new trim and lights, October '98
PROS Blindingly fast, tough as old boots, low-key looks
CONS Not for the inexperienced, often abused

CHECKING IT OUT

The Impreza Turbo comes as a tidy four-door saloon, or a five-door that's somewhere between hatch and estate. Prices are similar for both shapes. Both come as standard with power-assisted steering and anti-lock braking, driver's airbag, adjustable leather-trimmed steering wheel, electric windows and central locking. Later cars have a passenger airbag and a Thatcham 1 alarm/immobiliser as standard.

Go for a test drive and you should expect not just sensational performance from the 211bhp flat-four engine (up to 218bhp for the current model, with low-down torque boosted by an earlier tweak) but also a neat five-speed transmission, precise speed-sensitive steering and dependable brakes. Make sure the clutch has good feel and grip and has not been burnt out. It costs about £720 to fix.

The four-door featured in the photographs has an approved Prodrive interior with Recaro seats and air conditioning, which add about £500 to the value. The gold wheels fitted to some cars originally cost around £1500 (including tyres) and add a further £500 on a used car.

Check the small print of the car's warranty. New cars come

Five-door a genuine load-lugger

with a three-year/60,000-mile deal, but it is tied to a particular dealer.

If the car is labelled WRX, STi, or has two doors, then it is a grey import, except the 22B-STi Type UK, which is a full-blown rally replica built by Subaru Tecnica International Inc (hence STi) in a limited edition of only 16.

There is an approved McRae limited edition from 1995, individually numbered, with gold alloys and a sunroof. It was built by Prodrive, which provides all the approved performance and trim enhancements to UK Subarus.

The power and specification make some grey imports very tempting, but the prices are often high, and you must make sure the dealer selling you the car can back it up – Subaru UK dealers certainly won't, and that huge three-year official dealer guarantee means a lot.

There are some major differences between official and grey imports. The grey cars are often not undersealed, may not come with airbags or anti-lock brakes, and the suspension is of different construction.

Two-door body panels are not stocked by Subaru UK, and finding replacements for non-UK inner and outer trim can be a real drama. Histories are often uncertain, too. So be sure you know what you are letting yourself in for if you opt for an unofficial import.

With any Impreza Turbo, check for a complete service history, and note whether the 30,000-mile major service and 45,000-mile cambelt change are approaching.

Start inspecting the car with a thorough walk-round. Expect some light stone-chipping on bumper and bonnet, but also check the painted sills behind all wheels for damage from flying gravel. Look for dents picked up in car parks. The doors have side-impact bars but rather vulnerable skins and – until the '99 cars – no rubbing strip.

Feel under the front spoiler for grounding damage, and make sure the tyres are worn flat and evenly to show the suspension is straight. The Impreza Turbo is a very strong car, but heavy kerbing can distort the front wishbone inner support or bend the rear suspension links, and normal tyre wear at about £150 each will drop from a peak 40,000 miles to about 5000 with bent suspension. Around 20,000 miles per set is a realistic expectation for most drivers.

There should be no turbo screech when driving or smoke at idle. The turbo is well ▸

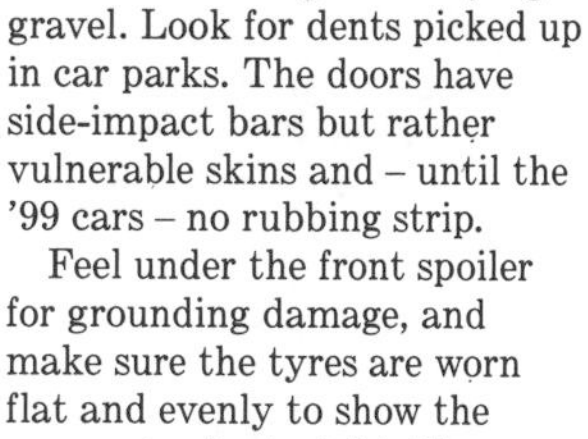

Most two-doors are grey imports

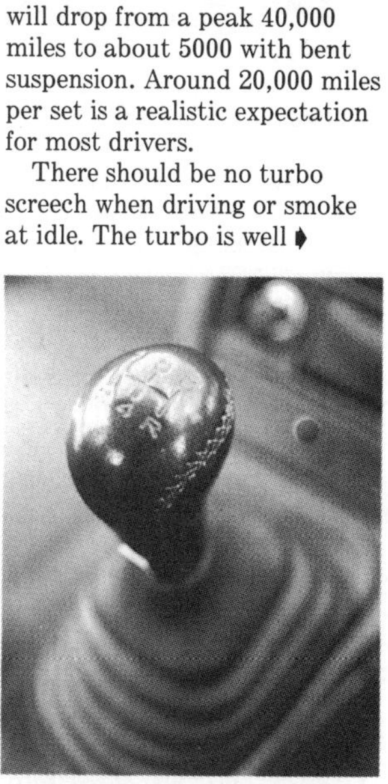

Five-speed gearshift is precise

Exhaust is well protected but check for damage, oil leaks, loose shielding

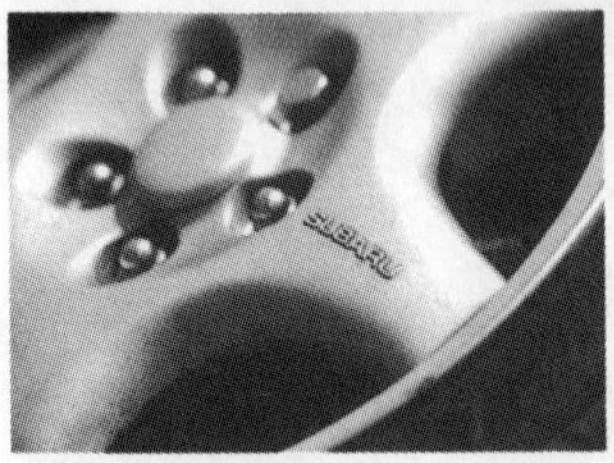

Inner front wishbone mount may be distorted if wheels have been kerbed

cooled from the bonnet duct, and should last the life of the car with correct service.

Under the car, the rally pedigree is obvious. The underside is very smooth, and the exhaust is protected by shields which crackle discreetly as the car cools down, and occasionally work loose.

A five-year life for the reasonably priced rear exhaust box is normal. There should be no oil leaks, and steering and driveshafts should be everlasting. Brake discs can take a beating, with quick drivers wearing them out every 20,000-30,000 miles, so check for grooving during your inspection and judder on your test drive.

This is a very strong and well-made car. So long as you insist on a thorough inspection inside, outside and underneath for damage and oil leaks, and provided the car drives well, there is little to catch you out.

Tough and honest with no nasty surprises ★★★★

WHAT HAPPENED WHEN

- **March '94** Four and five-door versions launched with 211bhp flat-four engine, five-speed transmission, four-wheel drive with rear and centre viscous differentials. Ventilated disc brakes with anti-lock, 15in alloy wheels, front and rear spoilers, bonnet scoop, immobiliser, central locking, driver's airbag, electric windows.
- **June '95** Prodrive-built Series McRae special edition launched with sunroof, gold alloys, low-profile tyres, individually numbered.
- **October '96** Facelift: lower bonnet line, new grille and bumpers.
- **October '97** New facia, twin airbags, 16in alloy wheels, sports seats, Momo steering wheel, engine uprated for low-down torque.
- **October '98** New interior trim and headlights, Thatcham 1 alarm and immobiliser, engine uprated to 218bhp.

Check for new or bent rear suspension links

Check bodywork for stone chips; front discs can wear out in 20,000 miles

Under-bonnet access fair; turbo should be long lived if serviced regularly

PARTS & SERVICING

Any Impreza Turbo you buy should have a full service history, and it's essential to maintain that to keep the car's resale value up.

A service every 7500 miles costs £102, and the major service at 30,000 miles costs £470. Bear this major service in mind when buying a car with mileage in the high 20,000s, and be aware that the £210 cambelt change takes place at 45,000 miles, and must be done by a Subaru dealer. The cambelt is very long and tricky to set up, and faulty fitment causes internal damage, or at least a serious loss of performance. The turbo should last for ever, but if you do need a replacement it will cost around £950. Destroy a clutch and that will be £720 fitted. Front discs, at £102 each, can go in about 20,000 miles.

PARTS COSTS	
● Front disc (vented)	£102.46
● Front pads	£88.66
● Rear disc	£99.59
● Rear pads	£71.59
● Headlight	£201.91
● Windscreen	£238.11
● Door mirror	£103.00
● Rear silencer	£162.93
● Front shock	£117.32
● Fuel filter	£36.22

Prices inc VAT but not fitting

Parts and service costs competitive by any standards ★★★★

INSURING IT

Like the Fiat Coupe and Nissan 200SX recently covered in Buying Used, the Impreza Turbo is surprisingly reasonable to insure. Cover from Rauch and Stallard in fact costs about the same for the Subaru as for those other two hot buys. If you have a clean licence, you'll be paying the same sort of premium as you would for the Golf 2.0 GTi, 24-valve Mondeo and BMW 325i. But because the Impreza Turbo is popular with knowledgeable low-life, most insurance companies will insist on a Thatcham 1 alarm/ immobiliser being fitted.

Supercar comparable with warm family saloon ★★★★

INSURANCE COSTS

96N, immobiliser, £250 excess

Age	Location	Fee
30	Guildford	£513
30	Matlock	£485
45	Guildford	£362
45	Matlock	£362

Official Subaru dealer such as Bell and Colvill best bet for peace of mind

Saloon has neat boot spoiler; interior has had slight tweaks over the years

WHAT THEY COST

To start on the Impreza trail realistically costs you 12 grand, whether you buy with a full warranty from a Subaru dealer, take a risk privately, or sort out a cast iron deal covering future spares supply with an established grey import dealer.

Total pre-purchase inspection is vital

Mileage is high on cheapest cars

Although reliability is high, you'll have to bear in mind that these £12k cars will typically have done around 90,000 miles. They will have been banged about a bit, and you could soon be facing bills totalling £1700 for a replacement clutch and turbo, which is why we'd always go with a Subaru dealer.

Go for a newer, lower mileage UK example in better condition with proven history for a few grand more and, if you look after it, you can expect devoted service for years to come.

And if you still want more power, Subaru dealers fit approved Prodrive 250bhp conversions with a warranty for a competitive £2450.

Mileage and condition crucial to price, residuals terrific ★★★★

PRICES WE HAVE SEEN

YEAR	MODEL	MILES	PRICE	SOURCE
● 95M	4dr	87k	£12,000	Dealer
● 96N	Prodrive 5dr, air	51k	£15,000	Private
● 96N	Prodrive 4dr, air	26k	£17,295	Dealer
● 96P	5dr	50k	£15,000	Dealer
● 97P	4dr	37k	£17,000	Trade
● 97R	5dr	21k	£19,000	Dealer
● 98R	5dr, 16in alloys	12k	£20,500	Dealer
● 98R	4dr	18k	£21,000	Dealer
● 98S	4dr	8k	£22,500	Dealer

Recaro seats only interior highlight

Four-wheel drive great in corners

Conventional styling does not fool enthusiasts – or well informed thieves

LIVING WITH ONE

Mechanical engineer and drummer Andrew Armstrong has covered 36,000 miles in his May '96 five-door, many of them done with teenage children or his drumkit on board. He finds the car perfect for an enthusiast with family responsibilities.

The Impreza is undemanding to live with, yet is the most surefooted car he has ever driven, with astounding reserves of performance available.

Troubles have been scored discs, replaced under warranty at 14,000 miles, a loose heatshield, and supermarket car park dents. He averages 25mpg, and has ordered a '99 version which has upgraded brakes and rubbing strips. He's selling his Porsche 911 2.7 Targa, as he never uses it now!

Widely admired; many owners come back for more ★★★★★

SNAP JUDGEMENT

The 96N, 26,000-mile, one-owner four-door saloon we drove cost £21,781 in '96. Basic price at that time was £17,000, and the fact that Bell and Colvill estimates its current value as £17,295 with a year's full parts and labour warranty demonstrates the desirability and short supply of these cars. The original purchase price was pushed up by its Prodrive Recaro interior, air con and a Thatcham 1 alarm. Apart from some minor stone chips, the car was immaculate. It was likeable and relaxed in traffic, while on Surrey's bumpy and slippery C-roads its performance showed why a "showroom" Group N Impreza Turbo won the Welsh Rally outright. We were sorry to give it back.

Class car has practicality and great performance ★★★★★

THANKS TO: Rauch and Stallard, insurance (01702 348261); Bell and Colvill, West Horsley, Surrey, parts and service information, facilities, photo car (01483 281000)

OTHER CARS IN THIS SERIES: Peugeot 205 GTi 4.9.96; VW Mk II Golf GTi 16.10.96; BMW 3-series 8.1.97; Range Rover 25.6.97; Jaguar XJ6 (1986-94) 3.9.97; Mercedes-Benz 190 12.11.97; VW Corrado 7.1.98; Fiat Cinquecento 4.2.98; Rover 200 18.3.98; BMW 5-series 10.6.98; Land Rover Discovery 5.8.98; Mazda MX-5 9.9.98; Peugeot 406 14.10.98; Fiat Coupe 11.11.98; Renault Clio (1991-98) 13.1.99; Nissan 200SX 17.3.99

PRODRIVE SUBARU IMPREZA TURBO RB5 LIMITED EDITION

Burning sensation

Here's a good investment tip. For £27,545, you can acquire a Prodrive Subaru Impreza Turbo RB5 Limited Edition, the car with possibly the longest name in the world. And if the Series McRae, Catalunya and Terzo editions are anything to go by, it will become a collectors' item. The 'RB' bit comes from the initials of Subaru's rally ace Richard Burns, after whom the car is named. The '5' is his racing car number. Alternatively, for £24,995 there's a non-Prodrive RB5 with the standard 215bhp engine. So now you know.

But if you know about collecting, you'll know items hold their value better when kept in the original packaging. Which means you'll lose out on what the RB5 has to offer. Because everything that matters to a driver, a *real* driver, has been worked on, from the firm ride to the feel of the handbrake. While trivia like styling gets a back seat.

First and probably foremost on the list of things that matter is the flat four engine. Start it up and blip the throttle. It growls. Find a stretch of open road and put your toe down. Get to 6,000rpm and that howl you can hear is the sound of 237bhp being developed.

By this time, of course, you'll be experiencing the kind of searing acceleration normally the preserve of decently powered motorbikes. And it just keeps accelerating. Suddenly, overtaking can be done in the time it takes to say 'Stop dithering'. Which can be a problem, because the RB5 makes you want to overtake everybody, all the time. Imagine how popular that would make you. Especially with Plod.

But let's imagine you're on a deserted, wonderful road, devoid of Gatsos, boys in blue, or old people in white Metros. A chance to press on. The engine redlines at 7,000rpm and whereas some gearboxes change ratios smoothly at low revs only to get clumsy at high engine speeds, the Prodrive version's short-throw 'box remains snickliy accurate throughout. It's almost a shame to reach fifth, since the fun and games happen anywhere from 3,500 revs upwards. That's when 258lb ft of torque makes its presence felt. To give you some idea, 50 to 70mph in top takes a mere seven and a half seconds.

For the first few bends you'll find yourself braking that little bit too early. It just can't be helped. Ventilated discs all round and four-piston calipers

Prodrive's 237bhp RB5 will no doubt cause a splash with rally fans, keen drivers and those who have a penchant for elegantly crafted and sensitively applied 'RB5' stickers

comprise a system that blends efficiency with feel. After that, you'll be carrying a bit more speed into the corners, feeling through the steering wheel and the seat of your pants for the permanent four-wheel drive's traction and grip. Which is amazing, aided and abetted by low-profile P-Zeros on 17-inch Speedline titanium-effect alloys.

The chassis has been set up to understeer if pushed, but it's predictable. I did get the inkling, though, that even if the car's limits are beyond the reach of mere mortals and exploitable only by the likes of Richard Burns himself, it could be quite snappy on the edge.

This is all academic, though, because it won't come out of the box, will it? So you'll just have to look at the dated, unlovely exterior and the basic, under-designed interior. The outer aspect is enlivened by the massive air intake on the bonnet and a huge wing on the bootlid that's liable to get you nicked. The front foglamps are protected by plastic covers adorned with RB5 motifs. These are just as liable to get nicked. Or fall off, as one of ours did.

Inside, the two front seats are bucket-style. They're firm, but not uncomfortable – even after a 200-plus mile drive. And they support all the way up to the shoulders, just the ticket for those crazy cornering speeds. An airbag-equipped Momo steering wheel, slightly too large for my own taste, comes as standard, as does an airbag-equipped plastic facia for the passenger.

No doubt each of the 444 RB5s manufactured all have a buyer's name against them. However, you could save yourself £6,595 by getting a normal Impreza Turbo. You'll only be missing 22bhp at most, and you may find that you can enjoy it more, more of the time. Then, when you did want to take it for a spin, you wouldn't have to worry about its collectors' item status □

Story Colin Ryan
Photography Mike Johnson

FACT FILE

Model	four-door super saloon
Engine	1994cc, flat four, dohc 16-valve turbo, 237bhp
Performance	0-60 in 5.2secs, 145mph
Price	£27,545
On sale in the UK	now
Rivals	Mitsubishi Evo 6, Audi S4, Honda Accord Type R, Volvo S40 T4
Likes	performance, handling and noise
Dislikes	looks, basic interior, and having to share the road network with everyone else

THE IMPREZA LEGACY

The twin-turbo Subaru Legacy RSK should be an automatic success, but it has a problem – its brother. Can the Legacy step out of the fraternal shadow of the awesome Impreza Turbo, which just got even *more* intimidating in the shape of the new, race-raw RB5?

Story by Mike Duff. Photography by Gus Gregory

RETURNING TO MY HOME TOWN IN LANCASHIRE RECENTLY, I found that things had changed: hell-raising mates were settling down with previously unsupposed girlfriends, the neighbours were extending their extensions, the cat had put on weight. And *everybody* was driving hot Subaru Imprezas.

I had travelled north in an official-spec, 2000 Turbo example. A fine tool for the 300-or-so miles, but also (if I'm honest) a hard-core posing machine. Except that, within 10 minutes of my arrival, I had seen three Imprezas with considerably longer boot badges than mine – one of which overtook me in what would be best described as a definitive manner. All bore the extra rear foglight of the grey import.

Impreza is becoming a cult, and a well-subscribed one at that. Young men are realising that, with a spot of sympathetic credit, their dream car is just about within reach. And hence the superabundance of Japo-spec Imprezas with endless, Scrabble-bag acronyms – STi, WRX, Version R and, of course, the ultimate *22B*. Add to this equation the number of companies prepared to further tune your highly tuned Impreza, and suddenly you realise that this is no longer just a car. It's a major social phenomenon.

And yet, it's been an enthusiasm that Subaru (UK) Ltd, official importers of the marque, has made precious-little capital from. Limited by import quotas and emissions standards, unwilling to use Single Vehicle Approval (except with the 16 22Bs), all it can do is flog official models and refer go-faster customers to rallying partners in grime, Prodrive.

Until, that is, the arrival of this: the all-new, all-official, 100 percent-warranted and guaranteed-ungrey Impreza RB5. OK, so colour-wise it *is* grey ('blue-steel metallic' disputes Subaru), but in terms of content, it qualifies. A large part is a cosmetic workover – wheels, tyres, and interior. But there is also the option (fortunately fitted to this example) of the Prodrive *performance pack*. Peak power goes up to 240bhp and torque to 258lb ft. Not quite the level of the twin-turbo, electronic-diff versions. But then, the fully guaranteed RB5 is unlikely to eat a gearbox a week, as some of its more extreme brethren are reputed to do. It was, apparently, intended as a tribute to new Subaru rally star, Richard Burns. The '5'? Oh, that's just the average crop of licence points from a gentle run in the country. Because it's seriously fast. As fast as the waves of highly tuned, grey-imported Imprezas? To find out, we drove it against a highly tuned, grey, erm… Legacy.

Not that the Impreza / Impreza angle wasn't considered, it's just that we really didn't want this degenerating into a wall-peeing contest. Getting close to the outer limits of even the coolest Impreza is a contest bound to end in tears.

Rather, think of this Legacy, a twin-turbo, 270bhp *B4 RSK*, as an alternative interpretation of the ultimate Subaru.

Underneath the butch bodykits and behind the alloy wheels, these cars are remarkably similar. Both have flat-four engines, turbocharged and intercooled. Both have permanent four-wheel drive and various clever differentials. And both are enormously fast. Yet, in terms of dynamics and style, they manage to be very different.

Take the recognition factor. In the three days I drove the Legacy, no-one looked at it twice except a farmer in Berkshire who asked me (having heard the particularly gruff and menacing exhaust note) whether it ran on diesel.

Whereas, during the same 72 hours of Impreza stewardship, I was being stopped by the police about once a day. While photographing around central London, we could actually see the security cameras tracking our every move. Inconspicuous the Impreza (most definitely) ain't. Big alloys, big spoiler and big noise. Revved from cold, it does a good impression of a Gatling Gun.

And of course, everybody had an opinion: One particularly charming van driver seemed to speak for the entire metropolis: 'Who d'ya fink y'are, Colin f*****g McRae?'

The chase for rush-hour lanespace was not an easy one in the RB5. Indeed, despite the enormity of the available thrust, countryside overtaking manoeuvres seemed more difficult, as well. There seems to be something about the lights and scoops and spoilers bearing down that triggers an instinctive, all-move-closer-together-and-wag-our-fingers reflex on the part of much of the motoring public. Indeed, in the spirit of fearless investigation, I measured the effect scientifically.

There is a particularly sleepy village which I regularly pass through, which provides a perfect sounding board for what middle Buckinghamshire thinks of a car. Without an official speed limit, one's velocity becomes a largely moral decision. Yet narrow roads, a school and 'Caution! Children!', 'Caution! Horses!' and 'Caution! Retired Colonels!' signs are strong motivations to slow right down.

Which, of course, I do. Passing in the Legacy at a hedonist 45mph, nobody even looked up from their warm beer aroun the cricket ground, and the agrarian harmony continued unbroken. But then, as I approached in the Impreza, (the raucous, offbeat exhaust note bouncing off walls and shaki windows), I could positively feel the curtains twitch. Sure enough, despite a scrupulously observed 29mph, at least o old gentleman looked up from his dog-walking to shake his head in sorrow at my conspicuous excess.

However, when you actually take this pair away from prying eyes and extend them somewhat, the dynamic

SPECIFICATIONS
SUBARU IMPREZA RB5
Price
£27,545 (Performance pack)
Engine
1994cc 16V turbocharged flat four
Power and torque
236bhp at 6000rpm
258lb ft at 3500rpm
Transmission
Five-speed manual, four-wheel drive
Max speed
145mph
0-62mph
5.2sec
Fuel consumption
27.3mpg

SPECIFICATIONS
SUBARU LEGACY B4 RSK
Price
£26,950
Engine
1994cc 16V twin-turbo flat four
Power and torque
270bhp at 6500rpm
253lb ft at 5000rpm
Transmission
Five-speed manual, four-wheel drive
Max speed
150mph
0-62mph
5.6sec
Fuel consumption
25.8mpg

differences do quickly become apparent. Same engine, almost the same transmission, but in terms of character the grey and official Subarus are at best distant cousins.

The Impreza is an enormously dangerous car. Not dangerous as in unstable, unpredictable or prone to bursting into flames. No, an Impreza will not kill you, it will get you to kill *yourself*. This car is, put simply, almost too good at what it does, too much for any mortal on (just about) any British road. A turbocharged violation of the laws of physics.

Driven fast, the Impreza feels like it's hoovering itself to the ground, impervious to corners and crests. It feels so normal and stable, that you drive faster to increase the risk quotient somewhat, to get the adrenaline coursing. And then, when the buzz of overtaking helicopters wears off, you notch it up some more. And all the time everything is stable and normal, and *CRUNCH*... for the only inevitability about an RB5 driven quickly is that if you believe its flattery for a moment, a very heavy impact will be your reward. Having scared myself poopless within about 30 seconds of entering the first derestricted backlane, I forced myself to drive around at normal car speeds 'to acclimatise'. And then I mentally marked all the enormous accidents I would otherwise have driven straight into – brown Rover 213 making an unindicated right turn, a flock of sheep on the road. Very salutary. The RB5 should flash up some message like 'You are not God, agree Y/N' on the windscreen at frequent intervals, which stops it if you respond incorrectly.

Whereas, by comparison, the Legacy is considerably closer to feeling like a normal car. The Nintendo effect just isn't here, thanks to the Japanese domestic-market settings – soft damping, over-light steering and a spongy brake pedal. Apparently, Porsche did some development work on the chassis, which I don't doubt, but it's still set up more for comfort than all-out, rubber-laying pace. Not that it can't do corners: the suspension digs in gamely and the four-wheel drive pulls you round fine. Indeed, find yourself on a typically British, low-to- zero-maintenance, corrugated road surface and the Legacy steams along quite happily when prudence and discomfort have already wound back the pace in the Impreza. Similarly, on motorways, the Impreza can be a dreary place – loud and tyre-boomy, reacting to unintended steering input and jarring over rumbly surfaces and cats' eyes. Whereas the Legacy, doing the same task, is softer and smoother. OK, so I'll stop making excuses and get to the point – the Impreza is quicker, and it's mostly down to the torque. Whereas, (once the Legacy's 200kg handicap has been accounted for), the power-to-weight ratios are almost identical, the Impreza has considerably more mid-range – 258lb ft from 3500rpm versus the Legacy's 253lb ft at a high 5000rpm. So, the B4 must be worked hard to get the most from the engine, while at reasonably low speeds there's enough pull in the RB5 for even the high gears.

Indeed, it's only if you really balls it up and drop out of the boost zone that the Legacy (with its twin turbos and bigger lungs) does better. Forget to change down in the Impreza (unlikely with the special, quick-throw shift, but just pretend) and you can find yourself trapped in a boostless limbo for surprisingly long periods, as the rotor struggles to gather momentum.

Subaru interiors have always been biased to utility rather than cutting-edge ergonomics, and this pair is no different. At least the Legacy looks vaguely respectable, though, with digital climate control and clear, workmanlike clocks. Although an improvement on the original, the Impreza's cabin is still fairly awful: black and plasticky with little thought on where to put switches. Embellishments include air-conditioning and suitably rallyish, black-and-blue, part-suede bucket seats, which unfortunately dig heavily into the rear legroom reserves. The Legacy has tons of room, but those looking for the ultimate in speed and space combined are more likely to opt for the estate version.

But which to buy and own? Pricing is sufficiently close to justify a fair comparison – £26,950 for the Legacy (including the £450 optional spoiler) and £27,545 for the Prodrive-engined RB5. A grey victory would undoubtedly be the story here, but the RB5 is so completely other-worldy that it won't happen. Subaru (UK) is back on top of the pile, with what can be fairly described as the best real-world Impreza you can buy today. So there. But don't forget the Legacy B4. Not the ideal poseur's car, but a full-sized, 270bhp saloon with room for four adults, a spacious boot, and an instantly forgettable profile. Bankrobbers of Britain, I believe we have the ultimate getaway car.

The Subaru Impreza 22B-STi gets under your skin. It worms and wiggles its way into your nervous system with an injection of almost 300 bhp. Sparks of energy arc through your twitching left hand as you flick the gears of its close-ratio 5-speed transmission. Your heart thumps in your chest hard enough for you to actually *see* the 4-point racing harness pump up and down ever so slightly, as if it were taking on a life of its own—its 3-in.-wide nylon weave constricting like a snake across your upper torso. You are slick with sweat.

The 22B infects your brain with visions of greatness. Dream-like images appear as wispy gauze before your eyes: Cresting a dusty hill somewhere in North Africa at 110 mph; cheering onlookers scrambling for the brush. You *are* rally champion Colin McRae. You're the kid with the new pair of Nikes, going faster, turning sharper and jumping higher than all your other friends. And that big toothy grin, the one untainted by the responsibility of adulthood, has resurfaced across your face. Hidden all these years and just waiting for the right moment. In the 22B, that moment has arrived. But surely this kind of fun comes at a price; perhaps you've made a pact with the devil.

Or maybe you just finished picking up a gallon of milk and some Ding-Dongs from 7-Eleven. If, that is, your local convenience store happens to be in Tokyo—Subaru sold its awd blue rocket only in Japan—and you were one of only 400 lucky souls to have plunked down 5 million yen for the chance to own the homologation special. The entire lot sold out in only three days, and none were destined for

Subaru Impreza 22B-STi

Dancing with the Devil

BY JAMES HOPE
PHOTOS BY GUY SPANGENBERG

our shores. But thanks to some hard work on the part of Subaru's North American contingent, one of them clandestinely arrived here—on a strictly temporary basis—not too long ago.

Almost entirely hand-built by Subaru Tecnica International, Fuji Heavy Industries' racing arm, the 22B is a testament to the fact that Subaru really can build a fun car; a rally car for the street capable of holding its own against some of the world's most revered sports cars. At 4.9 seconds to 60 mph, it's faster than a BMW M roadster or a Porsche 911 Carrera. It will go wheel-to-wheel with the Acura NSX and loses only one tenth of a second to the Corvette. That all this unadulterated speed emerges from a tiny 2.2-liter engine is no less than astounding.

Utilizing a closed-deck cylinder block surgically removed from Subaru's 1997 World Rally Car, the 22B's power is derived from larger-bore forged pistons originally found in the 2.0-liter horizontally opposed 4-cylinder WRX (*Open Road*, Summer 1996). Great gulps of air are fed through the gaping hood scoop and funneled into a huge aluminum intercooler. A button inside the cabin allows the driver to direct a cool stream of water across the intercooler's baffles to keep the turbocharger from incinerating itself.

The dohc 4-cylinder makes none of the noises normally heard from Subaru's gravelly sounding engines. A low-restriction exhaust blasts the high-strung engine's spent gases out a chrome-tipped tailpipe the diameter of a coffee can. With its soaring 7900-rpm redline and rather odd engine architecture the 22B comes off sounding like a Honda VTEC powerplant with a chest cold; low and *warbley* at low revs,

The 22B infects your brain with visions of greatness... cresting a dusty hill... at 110 mph

■ Seatbelt system isn't overdone. It's to keep you firmly planted when hurling the rice-rocket through its paces. Rotary knob locks the center diff, water spray button allows driver to prevent the intercooler from frying.

screaming like a tea kettle above 6500 rpm.

Japan's strict limitations on power output forced Subaru to affix a modest rating of 280 bhp at 6000 rpm to the car's technical specifications, but it's no secret that the 22B puts out closer to 320 bhp. The listed torque output of 268 lb-ft at 3200 rpm also seems conservative, with maximum punch probably coming in at closer to 280 lb-ft.

Those kinds of numbers require that the standard Impreza's clutch make way for one more suited to racing applications. In the 22B's case it's an ultra-light twin-plate ceramic/metal unit backed up by the transmission's five super-close ratios. The clutch is a robust piece that essentially works as an on/off switch to the 22B's front and rear halfshafts. Take-up is so minuscule as to cause lots of jittery starts and fits, but mastering the foot/hand coordination required to effortlessly fire off shift after shift is to ascend to automotive nirvana. Working the leather-covered ball of the shifter lever in the left-drive 22B is more of a thinking process than actually rowing through the gates. A mere flick of the wrist grabs the next gear with hardly any drop in revs. Those who have driven a Mazda Miata or Acura NSX know this ethereal sensation well.

Weighing in at a lithe 2799 lb makes the 22B somewhat light on its feet, and if it weren't for the car's full-time all-wheel-drive system, traction would be a considerable obstacle to overcome. But the mechanics of this Subaru's surefootedness are a bit different from those found in the grocery-getter Legacy wagon. A Variable Torque Distribution (VTD) system adjusts fore/aft power ratio through a planetary gearset located within the transmission. The ratios may be set from the cockpit via a rotary knob that controls the center differential from a "free" 35/65 front/rear split on up to a locked 50/50 power distribution. Four small indicator lights in the dash binnacle let the driver know exactly what percentage has been dialed in.

Like its engine and drivetrain, the 22B rides on a suspension setup with a heavily vested interest in racing performance. On-road ride comfort is not what this car is about. Divots and Tarmac imperfections go straight to your gut, spine and the base of your skull. But who cares when, on a whim, you can pull nearly 1.0g through a corner?

Long practiced in the motorcycle industry, inverted Bilstein struts increase the suspension's rigidity and contribute to more predictable handling characteristics. This arrangement increases unsprung weight, but Subaru balances this by using forged aluminum lower-control arms up front and super-light aluminum 17-in. BBS wheels, painted gold, of course, just like Colin's car. That they are wrapped in ultra low-profile 235/40ZR-17 Pirelli P-Zero tires speaks loudly about this car's performance pretensions.

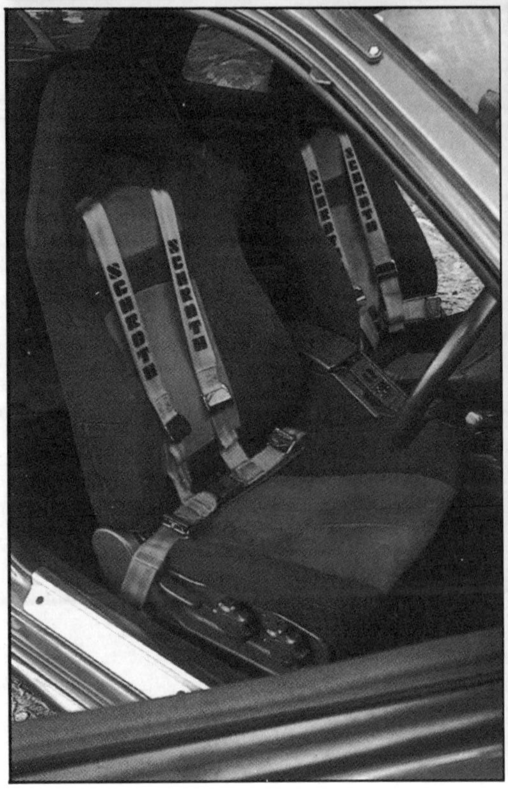

Subaru
Impreza 22B-STi

Subaru of America, Inc.
Subaru Plaza
P.O. Box 6000
Cherry Hill, NJ 08034
www.subaru.com

At a Glance...

0–60 mph	4.9 sec
Skidpad	0.93g
Slalom	61.1 mph
4wd system	full-time all-wheel drive
Max towing	na
Max cargo	11.1 cu ft
Max seating	5

List price: 5 million ¥ Price as tested: 5 million ¥

Price as tested includes standard equip (all-wheel drive w/driver-controlled center differential, power steering, a/c, tilt steering, 2-stage adjustable rear spoiler, sport bucket seats with suede inserts, 4-point racing harness, driver controlled intercooler spray, fender flares and side skirts, aluminum hood; power windows, mirrors & door locks).

SPECIFICATIONS

Engine

Type....**aluminum block and heads, H-4**
Valvetrain....**dohc 4-valve/cyl**
Displacement....**135 cu in./2212 cc**
Bore x stroke in./mm...**3.81 x 2.95/ 96.9 x 75.0**
Compression ratio....**8.0:1**
Horsepower (SAE)....**280 bhp @ 6000 rpm**
Bhp/liter....**126.6**
Torque....**268 lb-ft @ 3200 rpm**
Max engine speed....**7900 rpm**
Fuel injection....**elect. multi-point**
Fuel....**premium unleaded, 92 pump octane**

Chassis & Body

Layout....**front engine/awd**
Body/frame....**unit steel**

Brakes
Front....**11.5-in. vented discs**
Rear....**10.5-in. vented discs**
Assist type....**vacuum**
Total swept area....**339 sq in.**
Swept area/ton....**224 sq in.**

Wheels
cast alloy, 17x8.5

Tires
Pirelli P zero, 235/402R-17

Steering
rack & pinion, power assist
Overall ratio....**13.0:1**
Turns, lock to lock....**2.3**
Turning circle....**na**

Suspension
Front: **Inverted MacPherson struts, lower A-arms, coil springs, tube shocks, anti-roll bar**
Rear: **Inverted MacPherson struts, trailing arms, twin parallel links, coil springs, tube shocks, anti-roll bar**

Instrumentation

180-kph speedometer, 9000-rpm tach, coolant temp, fuel level, odometer

General Data

Curb weight....**2799 lb**
Weight dist (with driver), f/r, %....**57/43**
Wheelbase....**99.2 in.**
Track, f/r....**58.3 in./59.0 in.**
Length....**171.9 in.**
Width....**69.7 in.**
Height....**54.7 in.**
Minimum ground clearance....**5.5 in.**
Max approach angle....**15.0 deg**
Max departure angle....**16.0 deg**
Max towing capacity....**na**
Gross vehicle weight rating....**3406 lb**

Accommodations

Seating capacity....**5**
Airbag(s)....**none**
Head room, f/r....**est 39.0 in./36.5 in.**
Leg room, f/r....**est 42.5 in./32.6 in.**
Shoulder room, f/r....**est 53.0 in./52.4 in.**
Hip room, f/r....**est 51.0 in./52.2 in.**
Front seat travel....**8.5 in.**
Front seatback adj....**53.0 deg**
Cargo capacity, all seats up....**est 11.1 cu ft**
seats down, max....**na**
Cargo liftover height....**26.5 in.**
Max load floor length...**37.5 in.**
Width btwn wheelhouses....**41.0 in.**

Maintenance

Oil/filter change....**7.5 mo/7500 mi**
Tune-up....**est 30,000 mi**
Basic warranty....**36 mo/36,000 mi**

PERFORMANCE

Drivetrain

Transmission ....**5-speed manual**

Gear	Ratio	Overall ratio	Rpm	Mph
1st	3.08:1	13.7:1	7900	42
2nd	2.06:1	9.14:1	8000	65
3rd	1.54:1	6.83:1	7500	86
4th	1.15:1	5.10:1	est 7000	est 111
5th	0.82:1	3.64:1	est 7000	est 130

Final drive ratio **4.44:1**
Engine rpm at 60 mph in top gear......**3000**
Transfer case type ....**none**
Low range ratio ....**none**
Locking differential(s) ....**center**

Interior Noise

Idle in neutral....**51 dBA**
Maximum in 1st gear....**81 dBA**
Constant 50 mph....**74 dBA**
70 mph....**76 dBA**

Fuel Economy

Normal driving....**15.8 mpg**
EPA city/highway....**na**
Fuel capacity....**15.9 gal.**
Cruising range....**251 miles**

Acceleration

Time to speed....Seconds
0–30 mph....**1.7**
0–40 mph....**2.5**
0–50 mph....**3.8**
0–60 mph....**4.9**
0–70 mph....**6.7**
0–80 mph....**8.3**
0–90 mph....**10.7**
0–100 mph....**13.0**

Time to distance
0–100 ft....**2.7**
0–500 ft....**7.4**
0–900 ft....**10.6**
0–1320 ft (1/4 mi)....**13.6 @ 102.0**

Braking

Minimum stopping distance
From 60 mph....**157 ft**
From 80 mph....**277 ft**
Control....**good**
Brake feel....**excellent**
Overall brake rating....**average**

Handling

Lateral acceleration (200-ft skidpad)....**0.93g**
Balance....**heavy understeer**
Speed thru 700-ft slalom....**61.1 mph**
Balance....**heavy understeer**
Lateral seat support....**excellent**

TEST NOTES

- The 22B is a testament to Subaru's sports-car intentions with eye-popping acceleration and outstanding lateral grip.
- Overall ride comfort is seriously diminished with 22B's racing suspension and low-profile tires.
- The 5-speed gearbox is simply marvelous with seriously short throws and very close ratios.
- This is the ultimate rice rocket street racer—if only Subaru would bring it here.

PLUS

- Driver-controlled torque distribution
- Wild, aggressive good looks
- Perfect car to realize your rally fantasies

MINUS

- You can't get it here, no matter what
- Wrapping yourself around the seatbelts
- You still can't get it here

But these components whisper when compared to the 22B's overt styling. Done with about as much understatement as a Dennis Rodman hair scheme, the Sonic Blue Mica paint and wild bodywork scream for attention. Nevertheless, as this is essentially a race car, everything is done with a purpose. The enormous rear wing is manually adjustable up to 17 degrees for increased downforce. Bulging the fenders allowed for a wider track. The gaping maw of the imposing front air dam enhances airflow to the radiator. The list goes on and on. It's definitely a head turner, and no matter where we drove around rice-rocket-crazy Southern California, the 22B garnered admiring looks and up-turned thumbs.

Cruising, however, is not how the 22B is best enjoyed. It needs to be poked and prodded, examined closely on a dirt fire road or a twisting ribbon of asphalt. Its rubber squealing with delight through each apex. Your insides being pressed against your rib cage. Your sweaty hands guiding the three-spoke STi steering wheel with minute inputs. Really pushing the 22B rewards you with smile-inducing power-on four-wheel drifts. There seems never to be a lack of grip, the VTD system invisibly sending power to the front wheels and pulling the little blue streak out of yet another hairy situation. What fun!

As the pilot, you are ensconced within the comfort and support of a deeply bolstered racing bucket. Folding yourself around the safety harness is a chore, but once seated the 22B looks little different than a stock Impreza. There are white-faced gauges as opposed to the black-on-white dials found in the regular Impreza, but that's about it. In fact, despite its racing underpinnings, the 22B includes such luxury appointments as power windows and door

locks, even air conditioning. There's a blank panel where the stereo normally resides, but at least an antenna has already been pre-wired.

About the only drawback to the 22B is that it was never available in the States. But not to worry, Subaru has a few tricks up its sleeve. "We've finally been given the green light," said Subaru spokesman Alex Fedorak. Which means that the all-new Impreza, due out sometime around 2001, just may be a screamer. It's a sure bet the VTD system will show up—though it may lack the driver-controlled dial—after working its way down from the top-end models, and next year Subaru will introduce a new V-6 for its Legacy sedans and wagons. If the next Impreza ends up being based on the larger car's platform, the six-shooter would fit quite nicely. Hmmm...in the near future we may all have the pleasure of feeling something like the 22B worming and wriggling its way under our skin.

Unassuming looking maybe, but the little 135 cu-in. 4-banger puts out over 300 bhp which will zap you from 0–60 in 4.9 seconds. Multi-armed carbon-fiber strut-tower brace clamps to inverted dampers.

SUBARU IMPREZA P1

Road Test Number 4435

Prodrive's brief: The Definitive Impreza Turbo, crafted to suit UK roads. So no more grey imports, no more crude Japanese conversions, no more excuses. It's crunch time

MODEL TESTED P1 **LIST PRICE** £31,495 **TOP SPEED** 155mph
30-70MPH 4.8sec **0-60MPH** 4.7sec **60-0MPH** 2.5sec **MPG** 19.5
FOR Rock hard looks, performance, stunning ride quality **AGAINST** Weak brakes, numb steering

The wait is over. Welcome to the Subaru Impreza P1, the latest and most eagerly awaited rally-derived road burner for the year 2000.

For all the standard Impreza Turbo's undeniable talent, and the performance upgrades available from Prodrive, it has been difficult for Subaru UK and punters alike not to acknowledge yet faster variations of the car intended for Japanese consumption but shipped to the UK as grey imports.

Best of the lot is the WRX STi three-door: a 280bhp homologation special with uprated brakes and a stiffer bodyshell. Just the car then, for Subaru to import and sell in small numbers.

But that wasn't the aim of this exercise. Merely import an existing model and you end up competing with hundreds of independent car traders doing exactly the same thing. A unique UK flagship based on the WRX was needed. A car developed and type approved on our roads to be sold as a model in its own right through Subaru UK's regular dealerships

That's exactly what the P1 is. A machine created out of the WRX, but shorn of many rough edges to maximise its potential as everyday transport.

DESIGN & ENGINEERING

So sound is the blend of performance, refinement and value in the standard Impreza Turbo that we've always considered it to be a better road car than the WRX.

There was only one place for Subaru to turn for the expertise required to tame such a car: Prodrive. The P1 is intended to be the first in a succession of limited edition models developed by the Banbury-based outfit with more Scooby sense than anyone outside the Subaru factory itself.

The decision to type approve a new model was driven by both marketing and engineering needs. The execution of such a project has been an uphill struggle that exposes the vast chasm that exists between Japanese and European production regulations.

The biggest hurdle for Prodrive was modifying the 1994cc horizontally opposed turbo engine to comply with European noise and emissions requirements while retaining the standard WRX's 277bhp. Top of the list of changes was a reprogrammed electronic control unit designed and manufactured by Fuji. A new catalytic converter has been fitted, the standard twin catalysts being ditched for a larger single item.

All three-door WRXs have smoked rear side windows. They've been ditched for lightly tinted green glazing

A rear wiper has been fitted for UK roads. Good idea, but we don't like the exposed mechanism in the boot

P1 can't hide its roots. The exhaust back box has STi stamped on the side, as does the front strut brace

The result is a cleaner and quieter engine, but one that loses none of the WRX's outright poke. Maximum power is an identical 277bhp at 6500rpm, while peak torque stays at 253lb ft at 4000rpm.

Subaru UK has never sold the two-door Impreza, so the car needed European type approval. Months of legal wrangles and miles of red tape later, complete bodyshells that meet all European safety demands are now leaving the factory in Japan with a solid rear bulkhead for greater structural rigidity. Twin airbags are also standard.

So large is Subaru Technica International's (STi) range of springs and dampers that Prodrive tried no less than 22 variations. In standard spec the WRX is too harsh for UK roads, so the P1 retains the WRX's rear springs but uses softer front springs and uprated dampers all round to achieve a more compliant ride.

The key is revised front and rear geometry with particular attention being paid to tyre choice. Prodrive's competition relationship with Pirelli has helped the chassis engineers develop a car that uses 17 rather than 16in alloy wheels and lower profile 205/45 P-Zero tyres (the WRX uses 205/50s).

Dynamic excellence needed to be matched by a bespoke look. Peter Stevens of McLaren F1 fame restyled key areas of the body; the most striking aspect is the lack of gold wheels. Subaru has called time on such extravagance and the new OZ rims are gunmetal grey, in anticipation of the world rally car using the same for the rest of the season. The front and rear spoilers are unique to the P1 and offer aerodynamic advantages over the obvious added visual aggression. Tests show the P1 requires 3bhp less to maintain 100mph than the standard car.

Demand has been strong enough for Subaru to double the total production run from 500 to 1000 units. They are all sold.

Developed in the UK for the UK market ★★★★

HISTORY

Subaru UK is keen to play down any connection the P1 has with other Subarus, but eagle-eyed Scooby fans will spot that the bodyshell is from the latest WRX STi three-door. Prodrive has offered tuning kits for the Impreza for some time now, and in 1998 developed the 240bhp WR Prodrive. Later that year came the 22B, a wide-bodied monster we felt was too harsh for the road. Best of the lot was the RB5, a 240bhp WR Prodrive liveried to celebrate Richard Burns's continued rally success.

P1 sports Hella H4 driving lights instead of foglamps, which work brilliantly in cross-country conditions where Mitsubishi's Evo VI still reigns supreme

Prodrive has developed the P1 to have ultimate mix of familiar Impreza performance with new ride refinement

Louvres supply air to intercooler

Strut brace betrays P1's STi roots

PERFORMANCE AND BRAKES

Just a quick glance at the P1's specification tells you all you need to know about the car's performance. Anything that has 277bhp in a body weighing 1283kg is going to be swift. And the P1 does not disappoint.

Nothing gets off the line like a turbocharged four-wheel-drive car. Full revs, sidestep the clutch and the resultant burst of acceleration almost beggars belief. While you're looking for the locomotive that feels like it's just walloped you up the backside, the computer is trying hard to keep up with a string of figures that reads 0-30mph in 1.6sec, 0-60mph in 4.7sec and an effortless 12.4sec to 100mph.

Boost builds very quickly between 2000-3000rpm, at which point the benefits of a properly sorted forced-induction engine leap sharply into focus. From here the engine undergoes a delicious change of character. That friendly trademark Subaru warble rises an octave and then screams all the way to the 8200rpm limiter. The sheer energy with which it slices through the last 2000rpm is especially impressive for a unit that delivers its peak power at 6500rpm.

It's not all top end, though. Leave the P1 in fourth and it romps from 30-50mph in just 5.9sec, pulling off a similar trick between 50-70mph in top – it takes 6.8sec. Mighty impressive stuff from just 2.0 litres.

The gearshift – STi rather than Prodrive-derived – is also excellent and the ratios are well suited to UK roads. We can verify that 155mph maximum speed too, and the inherent stability provided by the body addenda at such speeds.

But once again the Mitsubishi Evo VI GSR proves to be the thorn in the Subaru's seemingly impregnable armour. Taking how brisk the P1 felt during our session at Millbrook, the Evo's 0-60mph and 0-100mph times of 4.4sec and 11.2sec seem outrageous. It's the same story in the gears too. The Mitsubishi shaves 0.9sec off the P1's fourth gear time and a whole second off its fifth gear time.

We can't repeat such praise for the braking system. Frankly, we expected more from the P1 – more pedal feel and more resistance to fade. Pushed to extremes – but not too often – they deliver the goods, hence our 2.5sec 60-0mph time. But ask too much and they respond with substantial fade and, in some cases, clouds of smoke.

Stunning performance, lacklustre brakes ★★★★

PERFORMANCE AND SPECIFICATIONS

ENGINE

Layout 4 cyls, 1994cc
Max power 277bhp at 6500rpm
Max torque 253lb ft at 4000rpm
Power to weight 215bhp per tonne
Torque to weight 197lb ft per tonne
Installation Front, transverse, four-wheel drive
Construction Alloy head and block
Bore/stroke 92.0/75.0mm
Valve gear 4 valves per cylinder, dohc
Compression ratio 8.2:1
Ignition and fuel Hitachi ignition, sequential fuel injection, Hitachi turbo with intercooler

GEARBOX

Type 5-speed manual
Ratios/mph per 1000rpm
1st 3.17/4.8 **2nd** 1.88/8.1 **3rd** 1.30/11.7
4th 0.97/16.2 **5th** 0.74/20.6
Final drive 4.44

MAXIMUM SPEEDS

5th 155mph/7500rpm **4th** 133/8200
3rd 96/8200 **2nd** 66/8200 **1st** 39/8200

ACCELERATION FROM REST

SURFACE DRY

True mph	sec	speedo mph
30	1.6	33
40	2.5	44
50	3.6	54
60	4.7	65
70	6.4	76
80	8.0	86
90	9.8	97
100	12.3	108

Standing qtr mile 13.9sec/105mph
Standing km 24.6sec/127mph
30-70mph through gears 4.8sec

ACCELERATION IN GEARS

mph	5th	4th	3rd	2nd
20-40	12.7	7.8	4.9	2.7
30-50	10.9	5.9	3.6	2.1
40-60	8.8	4.4	3.0	2.2
50-70	6.8	3.9	2.9	-
60-80	5.7	4.0	3.3	-
70-90	6.1	4.4	4.0	-
80-100	6.5	5.0	-	-

STEERING

Type Rack and pinion, power assisted
Turns lock to lock 2.7

CONTROLS IN DETAIL

1 Dials clear and attractive in daylight but difficult to read at night 2 Steering wheel only adjusts for rake, but just the right size and shape 3 Excellent seats are a gaudy mix of black, red and blue fabric, with some grey Alcantara thrown in 4 Stereo low down and fiddly to use 5 Air conditioning standard 6 European spec includes twin airbags

SUSPENSION

Front MacPherson struts, lower wishbones, coil springs, anti-roll bar
Rear MacPherson struts, coil springs, anti-roll bar

WHEELS & TYRES

Wheels 7JJx17in
Made of Alloy
Tyres 205/45 ZR 17 Pirelli P-Zero
Spare Full size

BRAKES

Front 280mm ventilated discs
Rear 245mm ventilated discs
Anti-lock Standard

BRAKES

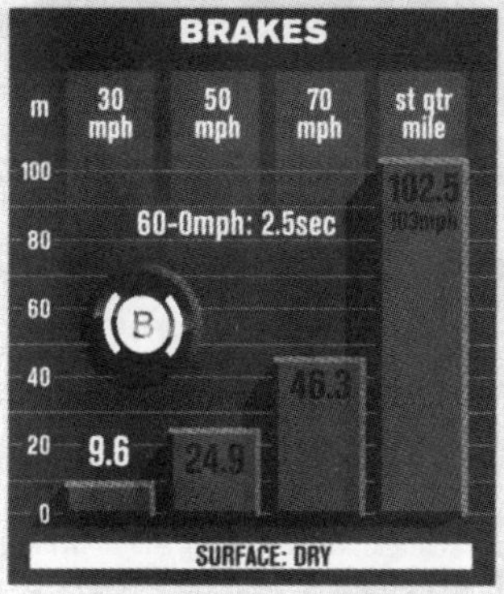

GEARING

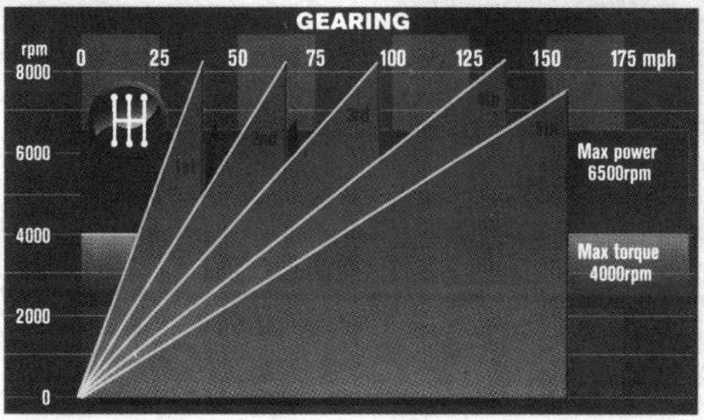

FUEL CONSUMPTION

TEST RESULTS

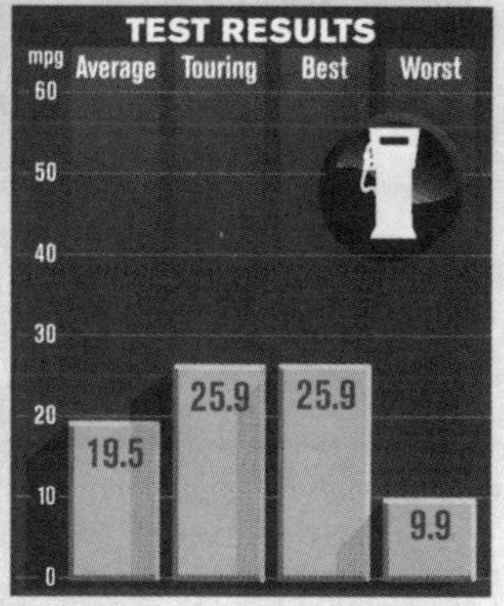

GOVERNMENT CLAIMS

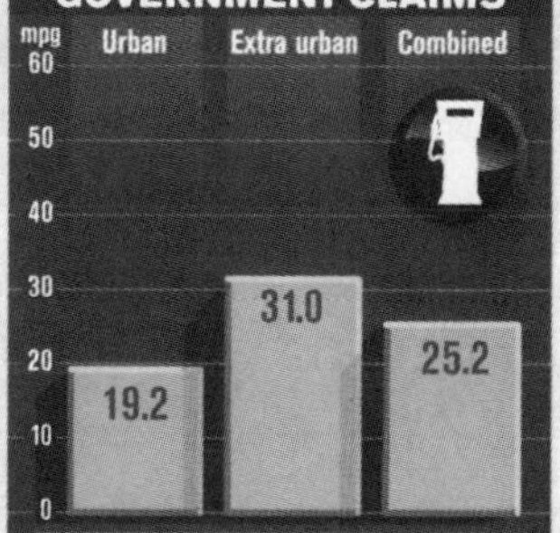

Tank capacity: 60 litres (13.2 gallons)
Touring range: 340 miles

NOISE

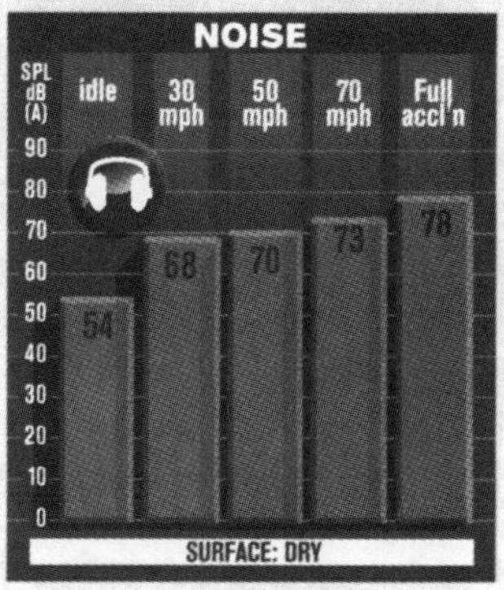

LAYOUT

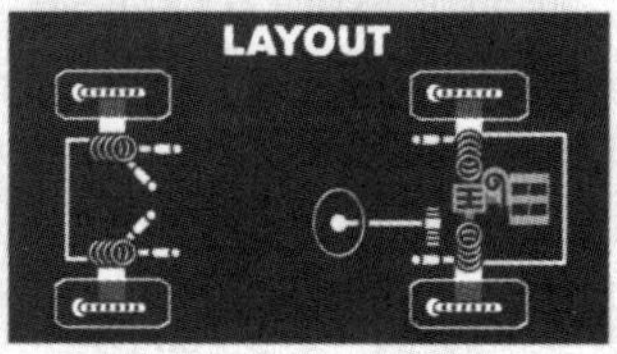

Body 2 door saloon **Cd** n/a **Front/rear tracks** 1459mm/1449mm **Turning circle** 10.4m **Min/max front leg room** 910/1110mm **Min/max rear leg room** 630/840mm **Min/max front head room** 930/980mm **Interior width front/rear**1410/1410mm **Min/max boot width** 960/1330mm **Boot length** 960mm **Max boot height** 380 mm **VDA boot volume** 365 litres/dm³ **Kerb weight** 1283kg **Weight distribution front/rear** 60/40 **Width (inc mirrors)** 1920mm

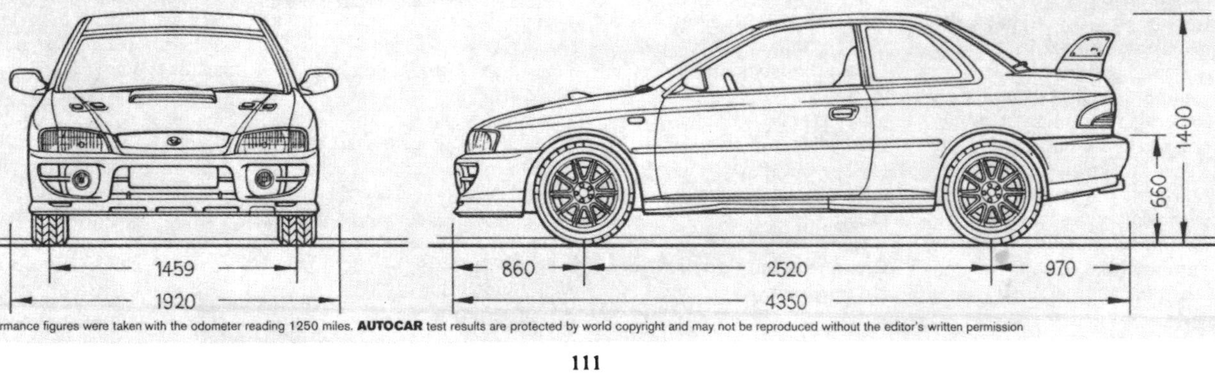

IAN HOWATSON

Ride and compliance are exemplary, with even the lumpiest roads tamed by the P1. But the pay-off comes in reduced driver satisfaction and feedback

Bodywork gets Prodrive stamp too

Scoops direct cooling air to brakes

Boot makes P1 a practical performer

HANDLING AND RIDE

Subaru had a clear objective in mind when it commissioned Prodrive to develop the P1. That it would have massive grip and traction went without saying. The critical ingredient deemed necessary for the British market was ride refinement.

No other Japanese high-performance car of this type comes close to matching it in this area. Trawling through town over potholes, the P1 feels as compliant as a BMW 3-series. Add some speed to the equation and it just gets better.

The P1 has a rare ability to smooth surfaces a driver's eye has marked down as being potentially troublesome. Under the same conditions a WRX would be an uncomfortable handful and even the standard Impreza Turbo wouldn't settle in the same manner.

But a handling and ride combination is a compromise, so something had to give. In the case of the P1, satisfaction and feedback have been forced to take a back seat.

Prodrive considered fitting a quicker steering rack to the car, but found the combination of that rack and the softer new front end made the car too nervous. So the P1 makes do with the standard Impreza rack, which takes 2.7 turns lock to lock, and loses a crucial degree of manoeuvrability. Challenging sections of B-road that should require nothing more than half a turn of lock in either direction need a couple of extra degrees. It's a small difference, but enough to mar the fun.

The other slight blot in the copybook is the P1's less than dogged resistance to understeer. We're used to Imprezas amazing us with awesome front-end grip, but the P1's front 205/45 ZR17 Pirelli P-Zeros start to slither earlier than expected. Perhaps the biggest disappointment is not being able to deal with the understeer by adding to or cutting the throttle.

What emerges is a chassis that is not the last word in driver involvement, rather one that's built for those who will use the P1 as everyday transport. What it loses to the Evo on a 50-mile blast over the Pennines, it recoups by dealing with all the M1 can throw at it in four hours. But the driving experience has suffered because of it.

Great ride, pity about the slow steering ★★★★

COMFORT, SAFETY AND EQUIPMENT

This is the area that has traditionally tended to throw the conventional supercar establishment into deep confusion. Japanese performance cars like this aren't just capable of destroying machinery costing twice the money cross-country, they can also offer ample space for four adults, as well as a large boot.

The recipe for the P1 is little different to any other Impreza *Autocar* has tested except for its lack of rear doors. The boot is exactly the same size as the four-door's. Front passengers get by far the best deal and sit in superb Prodrive bucket seats. It's difficult to think of road car chairs that offer the same combination of lateral support and long-distance comfort. The driving position is spot on, with the distance between the steering wheel and pedals proving comfortable for everyone who drove the car.

Had the P1 turned out to be a pared-to-the-bones racer, it might have been easier to excuse the lack of style and quality that pervades the entire cabin. The controls work well enough, but for £31,495 something more than the hard plastics that are barely palatable in a standard £20,000 Impreza are expected.

Here is where the P1's superb ride pays dividends. It really is a car in which long journeys are no hassle whatsoever. Wind and engine noise are kept well in check, and tyre and suspension thump are all but non-existent.

Standard equipment is typically generous, with electric windows, central locking, a Thatcham-approved alarm and immobiliser and air con all fitted as standard.

Safety is one of the P1's best assets. Both front seat occupants have an airbag, and anti-lock is standard. But it's the high levels of active safety inherent in a fast four-wheel-drive car that help the P1 score well. In our eyes, all that grip and power leaves it better qualified than most to avoid accidents in the first place.

Still practical, even as a two-door ★★★★

ECONOMY

The fuel figures start to look like they should come from a much larger engine than the P1's 1994cc if you keep your right foot to the carpet. At the track we recorded 9.9mpg, and our poor 19.5mpg test average included some sedate cruising.

Use a lighter right foot, and you might equal our 25.2mpg touring figure. Easier said than done once you've heard the engine at 7000rpm. At best, 342 miles is available between fill-ups of super unleaded.

Loves guzzling but can be surprisingly frugal ★★★★

Supremely comfortable bucket seats give occupants superb back support

THE AUTOCAR VERDICT

STAN PAPIOR

Body control sets hot saloon standards, but overall it's not a mould-breaker

MARKET AND FINANCE

Over the past five years the Impreza Turbo has built a brilliant reputation for reliability and rock-solid residuals. It was one of the few cars that worked as a rational as well as an emotional purchase.

The situation has weakened a little over the past six months but that won't affect the P1.

Each of the 1000 examples due to be produced has found a home and a reserve waiting list has been started.

Being an official Subaru model does wonders for the P1's reputation on the market. The £31,495 price isn't cheap, but the specification is high and no-one can argue with a three-year, 60,000-mile warranty.

Full UK back-up and they're all sold already ★★★★

WHAT IT COSTS

On-the-road price	£31,495
Total as tested	£31,495
EQUIPMENT	
(**bold** = options fitted to test car)	
Air conditioning	●
Traction control	n/a
Leather trim	n/a
Height/tilt-adjust steering column	n/a/●
Anti-lock brakes	●
Airbag driver/passenger	●/●
Satellite navigation	n/a
Alarm/immobiliser	●/●
RDS stereo/CD player	●/●
● standard - not available	
INSURANCE	
Group	20
Typical quote	£2100
WARRANTY	
36 months/60,000 miles, 6 years rust	
SERVICING	
Major 15,000 miles	
Minor 7500 miles	
PART PRICES	
Oil filter	n/a
Air filter	n/a
Set of spark plugs	n/a
Brake pads f/r	n/a
Exhaust (excluding cat)	n/a
Door mirror	n/a
Tyre	n/a
Windscreen	n/a
Headlamp unit	n/a
Front wing	n/a
Rear bumper	n/a

TESTER'S NOTES

Everywhere we went in Wales, people recognised the P1 as something a bit special. This side of the Severn bridge it just faded into the swelling ranks of hot Imprezas. It still looks tough enough for us, though.

The VDO stereo fitted to the test car deserves a special mention for being the fiddliest unit we have ever tried. No knobs, no preset numbers, and an array of warning beeps. What is the in-car hi-fi world coming to?

Rally ace Richard Burns won't have much time to pootle around in the company P1 – he's just bought a 911 GT3. Good though the Scooby may be, we'd take the Porsche on all but the wettest days.

The 205/45 Pirelli P-Zero tyres are an unusual size developed specially for P1, but they will soon be available for any car. For ride and refinement they are unbeatable.

IS THE P1 the ultimate road car, or even just the ultimate Impreza? The answer is not quite.

At its unveiling Subaru and Prodrive made no secret that this car was designed to fulfill two briefs. First, it was to be the best Impreza ever on the road. Second, it would wrest the cross-country crown from the fearsome Mitsubishi Evo VI.

In our view it succeeds on neither front. The P1 is definitely faster than the Turbo and the RB5. But it is questionable whether it's any more fun, and you can read over the page how we rate it against the Mitsubishi.

There's little point trying to hide our disappointment with the brakes and steering. It's a real pity the brakes weren't developed to match the rest of the car dynamically. As for the steering, it's a case of what might have been.

In isolation the P1 is a shatteringly impressive car, one whose body control and suspension composure set new hot saloon standards. But it is not the mould-breaker we had expected. Perhaps we expected too much.

Not the Impreza masterclass we expected ★★★★

Photography: Paul Harmer

Chasing the Dragon

Subaru's Impreza P1 mixes STI–developed grunt and a Prodrive developed chassis with devastating effect. Steve Bennett takes it to Wales for a spot of rally chasing

There's an Impreza P1 down in the car park. Do you A: polish and preen it before spending the weekend 'cruising' up and down your local high street basking in the admiring looks and cries of 'respect' from the hot hatch boys? Or do you B: load up with a few essentials and head to Wales to catch up on a bit of club rallying, while taking in some demanding roads in order to explore the potential of this, the ultimate Impreza (oh and bask in the admiring looks and cries of 'thrash it boyo' from the local rally boys).

The answer is, of course, fairly obvious, but on reflection I can't help thinking that sparing use of this car might not be a bad idea. The P1 you see is so devastatingly quick, so exploitable and so enjoyable, that holding on to your precious right to drive the highways and byways of this green and pleasant land maybe beyond your control. In short this car is a potential licence loser. But then it would be a shame to use it simply for, say, track days. It's dynamically too good for that. You need to experience its cross-country potential to really appreciate what a stunning, flattering car it is. So Wales it is then.

Andy Burton's Puegeot Cosworth being chased by Caterhams

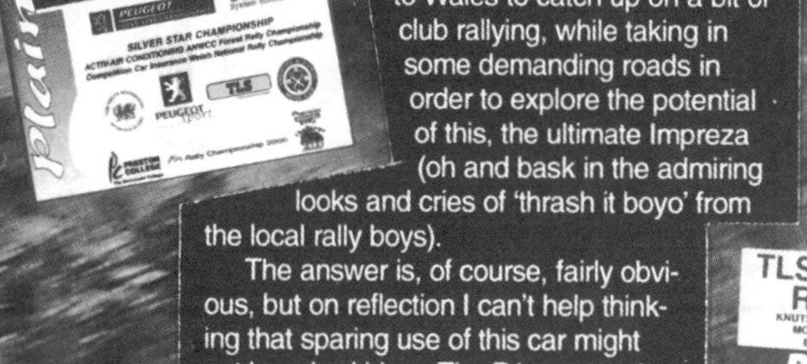

Before we get cracking, there is something we should get out of the way because we know that you're thinking it. Yes it is another Impreza. And yes you can't move for the bloody things and yes there are now magazines dedicated entirely to all things 'Jap sportscar' and other mags that should really change their names to 'Impreza World.' But, believe it or not, we haven't tested an Impreza since the 22B a couple of years ago. We have resisted the hype manfully, selflessly even.

But this UK Prodrive P1 is a car that deserves to be explored, not least because it has the potential to be the best Impreza yet. The potential even to be the best point-to-point drivers's car. Hell, it's got the potential to be the fastest f****** thing on four wheels. Sorry about the intimated swearing there, but there are occasions when only an expletive will do.

The event we're heading for is the Plains Rally, a round of the BTRDA Silver/Gold Star series; where club rallying meets forest stages and the breath of machinery stretches from Mk2 Escorts to Andy Burton's truly stunning mid-engined Cosworth powered Pug 306 spaceframe. The rally is run out of Newtown and encompasses stages in the Dyfnant and Dyfi Forest complexes, names synonymous with the Network Q Rally, albeit in somewhat warmer conditions.

Pointing the P1 in the direction of rally country, we leave London at 6.30am. The M4 proves that the Subaru can cut it as a motorway car; cruising almost silently, save for some tyre roar. Even the throbba-throbba off beat flat four beat seems to have been eliminated – such is the balance of the engine – which now produces 280bhp.

We head across country to Cheltenham before briefly hitting the M5 and then M50. We're off the beaten track now, ripping through Herefordshire. It's still early, the roads are quiet and the P1 is starting to reveal some of its astonishing ground covering ability. The build up is gentle. Confidence and boost levels start to increase. The P1 inspires. Its composure is rock solid and the short throw gear change blends with the solid power delivery. Corner sequences start to come together and sheer cornering grip and speed begins to take over.

Pro-driven

The P1 is Subaru's response to the grey trading of Scooby WRXs and the like, and the companies desire to create an Impreza to knock the Mitsubishi Evo 6 off its perch.

The engine had to be capable of 276bhp in order to take on the opposition. It also had to capable of passing the strict ETA noise and emission requirements, stricter than those in Japan. Intensive ECU development and an expensive single catalyst helped achieve the goals. Power is impressive enough but equally significant is the huge increase in torque. The P1 stumps up 253lb/ft at 4000rpm compared to the standard cars 214lb/ft. Top speed is an equally impressive 155mph with 0-60 coming up in 4.6secs.

While STI in Japan was responsible for the engine work, Prodrive concentrated on the chassis and styling. The former is what makes the P1 the car that it is. The 22B, for example, was way too stiff for UK roads, but for the P1 Prodrive's engineers have developed a handling package that is spot on. The anti-roll bars are the same as other Euro spec Impreza Turbos, but with firmer springs and dampers.

Those wings and things, as designed by Peter Stevens, have actually been developed in MIRAs wind tunnel. The result is the unique extra bib at the front and the carbon fibre (nice) gurney flap at the rear. The rear wing actually gives zero lift and saves 3bhp at 100mph.

Top: making a dust storm of our own. Right: ever the opportunist Bennett lines up to get his time card in

A trundle through Hereford town centre tempers things, which is almost a relief. Out the other side we pick up the pace and, once again, it's time to get your head around what the P1 is capable off. Firstly the P1 will go round corners like you would not believe and, while you're smack in the middle of that corner, you can feel what each tyre is doing as it balances grip against power. It feels great and it is that sort of information that makes you push harder and harder.

Grip is all very well and plenty of cars have it. What is just as important is ride quality and the ability to deal with surface changes and the sort of pock marks that blight our favourite driving roads. The P1 once again excels. It shrugs off bad roads, the Prodrive developed damping keeping the wheels on the tarmac rather than leaping and crashing around. This combination of grip and ride handling allows the P1 to get it's not inconsiderable power down like you wouldn't believe and contributes to an almost perfect balance of power, poise and grip that will encourage you to keep your foot hard in.

We arrive in Newtown. The rally has already started and the front running Silver Star cars are heading out of town to the first of the stages. There's something incongruous about competition cars running on the road, that and a sort of sense of occasion. The P1 picks up it's share of attention too. It looks different enough from the standard Impreza crowd with its Peter Steven's designed bodywork and

Astonishing levels of grip thanks to a chassis developed specifically for UK roads. Sheer hooligan cornering like this actually lays rubber down as wheels spin mid bend. Right: interior nothing special. Bit like the driver

unique OZ wheels. People seem to know what it is, and appreciate where it stands in the Impreza dynasty, which is probably at the top.

As rallies go this couldn't be much more compact taking in just four stages, starting at a civilised 10am and finishing at 2.45pm for the front running Silverstar cars. The Goldstar boys were running around an hour behind so they had even more time in bed. Rally HQ is based a few miles out of town and we pop in to collect our passes and route map, although with such a relaxed schedule, racing from stage to stage in Rally GB style isn't an issue. Shame really!

We head the rally off at the first and only service halt and marvel at the sort of organisation and volunteer spirit that rallying at this level demands. Sure the Ford Ka Championship and Daihatsu Avanzanto competitors have the luxury of hospitality of sorts, but for most it's a snatched snack while helping the mechanics change the wheels, refuel, bash out a wing or replace a bent steering rack. And for mechanics, read an assorted bunch of one's mates/family.

We're keen to see Andy Burton arrive in that amazing, aforementioned Peugeot. It is without doubt one the most impressive self-built machines we've ever seen. As it trickles into the service bay it is swamped. The entire rear deck lifts to reveal the spaceframe which bolts directly onto the central tub. Lurking inside is the tiny ex-DTM V6 Cosworth motor, the same unit which powers the Gould Hillclimb car in this month's Tech Focus.

Andy paces around the car, balancing mechanical duties while tackling an ice lolly. The brakes have been fading ,so the fluid is topped up and air from the system is bled with the help of Andy's wife on pedal pushing duty. A bent steering rack is replaced and a spanner check is conducted. Co driver Rob Morgan changes the air filter, which is choked with dust after just two stages. Otherwise it's routine stuff and Andy is looking good for an outright win.

To see this car in action is something else. After the service halt we trek off to SS4. Standing at a fast left/right wiggle at the top of a long uphill stretch, we can hear the wailing V6 on the other side of the forest. It

TECH SPEC
Impreza P1

Engine: Flat four 16v, turbo

Max power: 276bhp @ 6500rpm

Max torque: 260lb/ft @ 6500rpm

Weight: 2723lb/1235kg

Power to weight: 214bhp/ton

Transmission: Five-speed manual, 4WD

Wheels: 7x15 OZ alloy

Tyres: 205/45 ZR17 Prelli P Zero

0-60mph: 4.6secs

Top speed: 155mph

Price: £31,000

sounds more like a single seater, hardly surprising given that the tiny 2.5-litre V6 revs to over 11 000rpm. Trailing a Hiroshima-sized dust cloud, Andy flicks the Pug into the corner, applies a dab of lock and nails the throttle. The traction is awesome and you can visibly see the car dig in and power out. Stunning and a real crowd pleaser.

"It's possibly the maddest drive I've ever had. Yes I feel reckless and yes I feel positively dangerous"

Back in the Scooby and we indulge in a bit of dust creation of our own as we head out of the stage on the fast, compacted forest roads. Time to play again too as the P1 just wants to be driven faster and faster. The defining P1 moment occurs on the way back to rally HQ on a stretch of road that shall remain nameless. The guy on the Triumph T595 made it perfectly clear that he wanted to play by hanging back and waiting for us to clear a few straggling cars. Then he came over all Carl Fogarty like and nailed it.

Superbikes usually make most supercars look pretty sad. Sure they're not so quick through the corners but sheer ballistic speed usually makes up for it on the straights. Not today, though, and not with the P1. This guy was good and boy was he trying. Watching him decking it through the corners, knee scraping the black stuff, rear tyre squirming under power, was like one of those stunning televised on bike shots with the Impreza windscreen acting as the TV.

Once again the P1s uncanny ability to cover ground fast comes into its own. It has the sheer grunt to stay with the bike on the straights simply because it has managed to put the power down so much more effectively out of the corners. And, of course, while in the corner it is carrying a phenomenal amount of speed. And don't think for one moment that because I'm driving a Subaru that I must have suddenly been blessed with the driving talents of Richard Burns. No, the P1 is incredibly capable and very flattering to drive. You can use just about all the performance without any danger of frightening yourself stupid.

Superbike man eventually waves us through. It's a bit of relief and we wind the pace down a tad. It's possibly the maddest drive I've ever had. Yes I feel reckless and yes I feel positively dangerous. More than anything, though, I'm totally dumfounded at the ability of the P1. Nothing I have ever driven comes close to the point-to-point ability of this car. It may not be the most powerful of the Jap wonder wagons, but it uses what it's got to much better effect.

If there is a criticism it is with the brakes. They really aren't powerful enough or confidence inspiring. They can't quite keep up with the rest of the car. Other than that this is the ultimate Impreza and worth every penny of the asking price. Supercars, superbikes, super-bloody-anything; be afraid, be very afraid ■

It's all in the details. Carbon/alloy strut brace looks the part and looks beefy enough to work. OZ alloys designed by Peter Stevens

Only service halt on the event and the Burton Pug is in need of a new steering rack

SUBARU IMPREZA P1 WR

Magnum P1

A range of Prodrive upgrades has turned the already awesome Impreza P1 into a simply unbeatable World Rally Car clone

HAVE YOU HAD A LOOK AT THE pictures of the new Impreza Turbo elsewhere in this issue? Oh dear. It might be the dog's bollocks to drive, but what have they done with the styling? Makes you appreciate the current car even more, and especially the two-door P1, which combines the born-to-be-bad look of Subaru's World Rally Car with the best bits from the STI parts bin and some magic from Prodrive. Making it the best Impreza yet. Possibly ever.

Stimulating though it is to drive, the UK-only P1 has a couple of flaws that are just a little disappointing in a car with this much potential for going insanely quickly across country. One: the STI brakes don't seem up to the job of reining in a P1 on a mission. Two: the steering is a little slow-witted in tighter corners. Nothing to lose much sleep over, but imagine how awesome the P1 would be if things like that were sorted.

Enter Prodrive again. Just to refresh your memory, the Banbury-based team has done some great aftermarket things to Imprezas in the past, such as 1995's Series McRae, and the same is true this time with the P1 WR. Upgraded brakes, bigger wheels and tyres, a naughty exhaust system, new Recaro seats and gas discharge lights can all be fitted by any Subaru dealer, polishing off the P1's rough edges and propelling it even further into supercar-bashing territory.

By far the biggest leap in performance is made by the brakes, which cost a hefty £1527. Developed in conjunction with Alcon, the new front brakes are similar to those on the rear of the Subaru rally car – pukka 330mm diameter spiral-vented discs, four-pot Alcon calipers – but no small amount of work has been carried out to come up with the most suitable combination of materials for the road. In their latest evolution, the brakes feel like they've come straight out of a pukka racing car, with a relatively hard pedal that stands the car on its nose in a heartbeat with just a short stab and tests front-end grip to the full. They even squeak a little, although there's no need to warm them up before they reach full performance. There's so much feel that accurately judging the level of retardation soon becomes instinctive. It's unlikely you'll find more powerful brakes on any road car.

The driving experience is further improved by fitting the bigger wheels and tyres (£1980) and modifying the suspension geometry front and rear to match. The OZ Racing rims are the same as those of the standard P1, but an inch larger in diameter at 18 inches, and they're fitted with 225/35 ZR18 Pirelli P Zero tyres. Not only do these changes sharpen up the steering's responses but they also make the helm feel calmer and less prone to tramlining. Front-end grip runs out earlier than you might expect in tight or greasy corners, but Prodrive says it steered away from an aggressive set-up in order to make the car user-friendly at its sky-high limits.

The big-bore exhaust is also well worth £350, making the Impreza's charismatic flat four feel smoother and more free-revving than that of the standard P1 and pumping up the warbling exhaust note to a wickedly fruity level. And if you do a lot of night driving, the £470 Hella gas discharge spotlights light up the road like an alien spacecraft hell bent on a close encounter. Only the £1995 Recaro seats are questionable, if only because the standard P1's seats already do a fine job.

Add up the WR options package and it comes to £6322, on top of the £31,495 you will have paid for a P1. Make up your own mind if that's expensive or not; all I can say is that, even by normal Impreza standards, the time between points A and B on the map just got a whole lot shorter.
ALLAN MUIR

No changes inside but Recaro seats an option

Revised geometry, bigger wheels aid steering

Point to point, there's little that will beat P1 WR

Colossal Alcon front brakes stop you dead

Rating	★★★★★
Price	£37,817
Engine	1994cc dohc turbo flat four, 280bhp, 214lb ft
Performance	155mph, 4.7sec 0-60mph, 25.2mpg
On sale in UK	Now

ROAD TEST NUMBER 4464

SUBARU IMPREZA

The hotly anticipated new Impreza WRX faces the daunting task of replacing a stunningly effective driver's car

MODEL TESTED WRX **LIST PRICE** £21,495 **TOP SPEED** 141mph
30-70MPH 5.8sec **0-60MPH** 5.7sec **60-0MPH** 2.8sec **MPG** 18.7
Brilliant chassis and steering, refinement, value **AGAINST** Looks, poor stereo, economy

Ten years ago it would have been inconceivable for a Subaru to be the most eagerly awaited performance car of the year. Not now. The new £21,495 Impreza WRX is just that car.

Since 1994 the Impreza Turbo has defined affordable performance motoring, and in the process facilitated one of the most remarkable image rethinks ever witnessed by the motor industry.

And now the Impreza has to start all over again.

You'd think that all the hard work had been done over the last seven years, but the truth is, the arrival of the new Impreza is a pivotal moment not only for the Impreza but for the company itself. In many ways it's more difficult second time around: people's expectations are far higher now.

The theme is wholly familiar: a turbocharged boxer four-cylinder engine sits longitudinally in an all-wheel-drive chassis. But this time there's a greater concession to drivetrain and chassis refinement, as well as a new name. What was the good old Turbo becomes the WRX. And Subaru UK has confirmed that in 2002 the 215bhp car will be joined by a 280bhp model, possibly badged STi.

DESIGN & ENGINEERING

Let's deal with the main issue first: the looks. First off, rest assured that the new Impreza looks miles better in the flesh than it does in photographs. And being the WRX model it has arches with enough girth to make at least some sort of style statement.

But despite Subaru's claims that it was trying to create a shape that would improve over time and not date too quickly, the new Impreza WRX's styling is still a big disappointment. Park it next to the last of the old Turbos and the new car looks like a cruelly distorted caricature of its predecessor.

There's a lot that's carried over, though. The trademark front driving lights remain, as do the huge air intake on the bonnet and twin exhausts. The flared wheel arches are new and bear more than a passing resemblance to those found on another rally champ, the original Audi Quattro.

On balance it's a shape that's best in profile, at its weakest from the rear three-quarter and not very successful overall.

The rest of the car still makes mouthwatering reading. At the business end sits Subaru's familiar horizontally opposed four-cylinder turbocharged engine. It retains the same

1994cc capacity and boasts the same 215bhp power output as before. Peak torque rises by just 1lb ft to 215lb ft at 3600rpm, 400rpm lower than before.

There's no clear reason for the increase, but a new catalyst and an 11 per cent larger intercooler both make some difference. There's nothing wrong with using the same inherently excellent engine, but the power-to-weight ratio has been hit to the tune of 19bhp per tonne now that the car weighs 1385kg according to our scales, an increase of 150kg.

The real work has gone on around the bodyshell and in fine tuning the already excellent chassis. Torsional rigidity was the main goal and Subaru has achieved a whopping 250 per cent increase over the old car in saloon form. That leaves a much more solid platform on which the all-round MacPherson strut suspension can do its stuff. New pick-up points for the suspension, a redesigned front subframe, revised spring and damper ratings and a 33mm rise in the overall roll centre of the car should make quite a difference to the driving experience. The focus was on extra refinement and added adjustability, according to Subaru.

Tyre and wheel diameters rise to 17in, with 215/45 tyres all round. The wheels are an open-spoke design and leave a clear view of the ventilated discs (294mm front, 266mm rear) fitted as standard. There's no traction control, just a limited-slip differential at the back.

Big changes have been made to the interior, the one area where the old car attracted real criticism. An all-new dashboard and centre console houses nothing more imaginative than the usual stereo and rotary heater controls. It's difficult not to think that rather more than this was needed to tempt people away from more stylish European rivals.

Evolution of a proven formula
★★★★

HISTORY

It all started back in 1994. By whipping the turbocharged four-cylinder boxer engine out of its Legacy Turbo, adding a little more power and then slotting it into the smaller Impreza, Subaru created the car everyone wanted, the Impreza Turbo. In Japan it spawned all manner of special versions, from the cooking WRX to the WRX STi VI. It came here as the good old Turbo, with a few modifications along the way. Britain had two specials: the P1, which was the fastest, and the RB5, which was the best.

Acceleration and top speed suffer in comparison with the outgoing Impreza Turbo, but the new WRX is the better all-round high-performance package

Roomier cabin falls short of German quality, and wheel still lacks reach adjustment. Driving position is excellent

High lip lets down usefully big boot

Engine given only minor changes

PERFORMANCE AND BRAKES

What a relief. Subaru may be trying to add some deportment to the Impreza's CV by making it quieter and more refined, but it still had the foresight to make the WRX get off the line with plenty of drama.

Given maximum revs and a side-stepped clutch, the new WRX catapults itself from 0-30mph in just 2.0sec. One extremely swift gearchange later and 60mph arrives in just 5.7sec. Sustain the tempo and 100mph whizzes by in 16.9sec, the standing quarter mile in 14.5sec. Top speed at a rather blustery high-speed bowl was 141mph.

In isolation that's a very impressive set of figures, one that would take 231bhp of BMW 330i to beat. But there are a couple of snags. First, the figures show it to be considerably slower than the old car against the clock (that managed 5.3sec to 60mph and 14.9sec to 100mph, and did 143mph flat out). Second, the subjective impressions support such numerical shortfalls.

No matter what gear you're in or what speed you're travelling at, the Impreza no longer hurls you up the road in the brutal manner we've come to expect. Throttle response is good rather than exceptional for a turbo motor, with boost coming in as low as 2500rpm. But the mid-range isn't as strong as before and even though the rev limiter is set at 7400rpm, anything beyond 6500rpm and the performance starts to tail away.

Again, the in-gear times are impressive in isolation but well down on before. In fifth gear it manages the 50-70mph drag in 9.6sec, up from 8.5sec. And from 30-50mph in fourth it needs 8.0sec, a disappointing 2.0sec slower than before.

But that's where the bad news stops. This Impreza strikes a brilliant balance between refinement and character and makes you wonder why anyone bothers making in-line fours when they're so much rougher than the Impreza's boxer. It may not be as fast as the old car, but with the extra refinement it's almost as satisfying.

The gearshift and clutch are markedly improved too. There's less resistance and still a short, snappy throw.

Brakes have long been an Impreza strong point, and these maintain that tradition. Pedal feel is better than before; they didn't fade once, and they gave every driver real confidence.

More refinement, better gearbox, but less pace

★★★★

PERFORMANCE AND SPECIFICATIONS

ENGINE

Layout 4 cyls horizontally opposed, 1994cc
Max power 215bhp at 5600rpm
Max torque 215lb ft at 3600rpm
Specific output 108bhp per litre
Power to weight 155bhp per tonne
Torque to weight 155lb ft per tonne
Installation Front, longitudinal, four-wheel drive
Construction Alloy heads and block
Bore/stroke 92.0/75.0mm
Valve gear 4 valves per cylinder, dohc per bank
Compression ratio 8.0:1
Ignition and fuel Electronic ignition, sequential fuel injection, IHI turbocharger with intercooler

GEARBOX

Type 5-speed manual
Ratios/mph per 1000rpm
1st 3.45/5.3 **2nd** 1.95/9.4 **3rd** 1.37/13.4
4th 0.97/18.8 **5th** 0.74/24.7
Final drive 3.90

MAXIMUM SPEEDS

5th 141mph/5710rpm **4th** 132/7000
3rd 94/7000 **2nd** 66/7000 **1st** 37/7000

ACCELERATION FROM REST

SURFACE DRY

True mph	sec	speedo mph
30	2.0	31
40	3.1	41
50	4.3	52
60	5.7	62
70	7.8	72
80	9.9	83
90	12.8	94
100	16.9	104

Standing qtr mile 14.5sec/95mph
Standing km 27.9sec/120mph
30-70mph through gears 5.8sec

ACCELERATION IN GEARS

mph	5th	4th	3rd	2nd
20-40	16.3	9.6	5.6	3.2
30-50	14.5	8.0	4.3	2.6
40-60	12.2	6.4	3.6	2.6
50-70	9.6	5.3	3.6	-
60-80	8.6	5.6	3.8	-
70-90	9.4	6.0	4.6	-
80-100	10.3	6.6	-	-

STEERING

Type Rack and pinion, power assisted
Turns lock to lock 2.8

CONTROLS IN DETAIL

1 Dials are clear and easy to read, and look even better when they glow at night 2 Alloy-faced pedals are well spaced 3 Car seats don't come much better than this: comfortable, supportive and stylish too 4 Air conditioning now comes as standard 5 Poor-quality hi-fi sounds terrible 6 Driver and passenger airbags both come as standard

SUSPENSION

Front MacPherson struts, transverse link, coil springs, anti-roll bar
Rear MacPherson struts, trailing arm, dual link, coil springs, anti-roll bar

WHEELS & TYRES

Wheels 7x17in
Made of Alloy
Tyres 215/45 ZR17 Bridgestone Potenza RE 011
Spare Space saver

BRAKES

Front 294mm ventilated discs
Rear 266mm ventilated discs
Anti-lock Standard

BRAKES

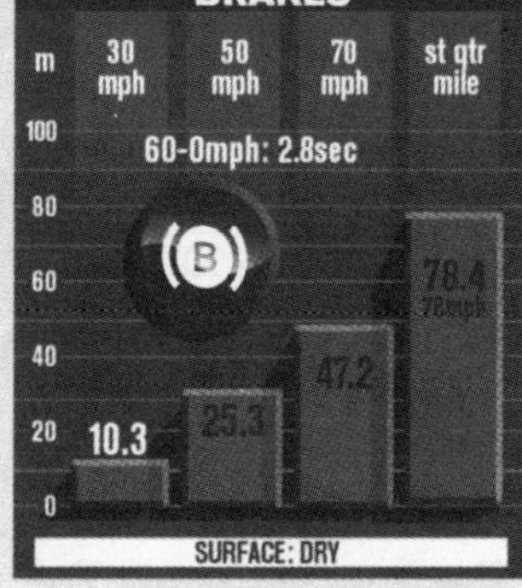

GEARING

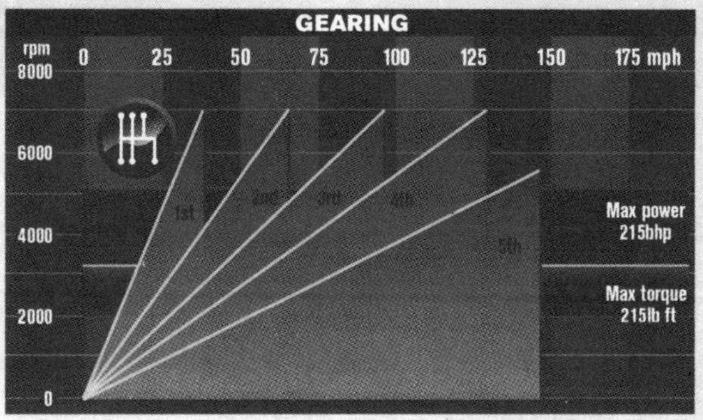

FUEL CONSUMPTION

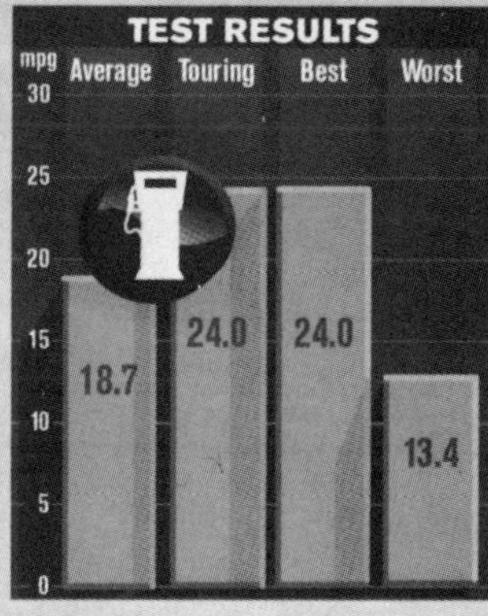

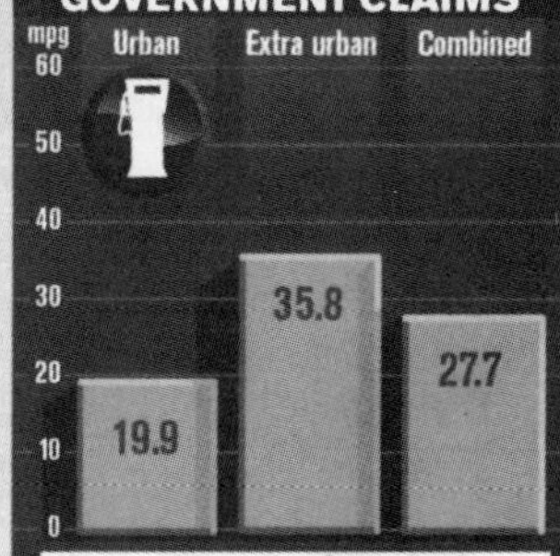

Tank capacity: 60 litres (13.2 gallons)
Touring range: 320 miles

NOISE

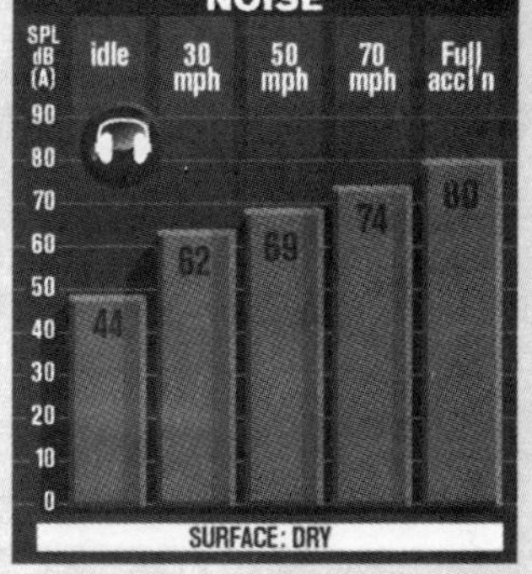

LAYOUT

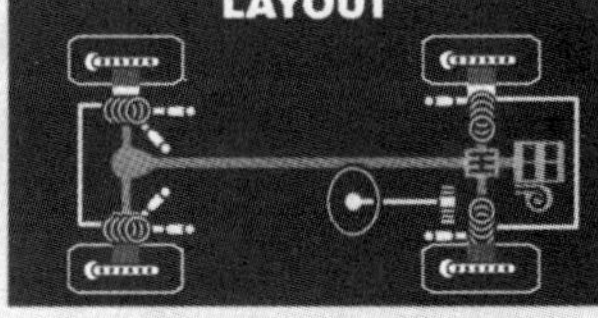

Body 2dr saloon **Cd** n/a **Front/rear tracks** 1485mm/1480mm **Turning circle** 11.0m **Min/max front leg room** 920/1120mm **Min/max rear leg room** 640/850mm **Min/max front head room** 920/1110mm **Interior width front/rear** 1430/1430mm **Min/max boot width** 960/1340mm **Boot length** 970mm **Max boot height** 380mm **VDA boot volume** n/a **Kerb weight** 1385kg **Weight distribution front/rear** 60/40 **Width (inc mirrors)** 1730mm

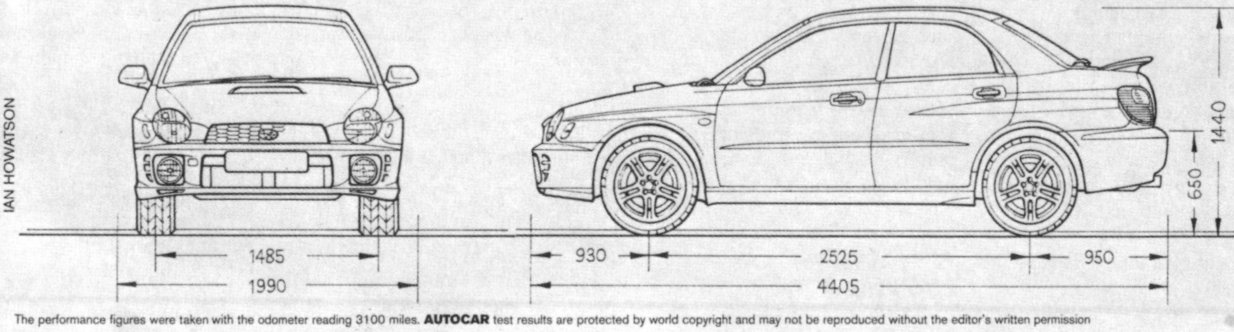

Get used to the revised steering and it proves to be a brilliant compromise that's ideal for both vigorous cornering and sedate cruising. Great grip too

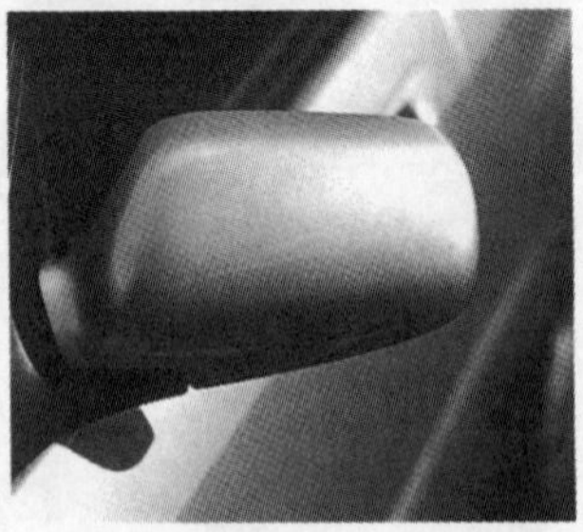

Turbo name has been replaced by WRX tag. Few options are available, but standard kit includes 17in alloy wheels

HANDLING AND RIDE

Where do you start with a car whose overall ability might just rewrite the manual on performance saloon handling? With the bits that are different.

Subaru seems to have taken the concept of comfort to heart, because in a stroke it has turned the Impreza from being something to avoid over long distances into a comfortable saloon. New spring and damper ratings all round have given the suspension a compliance totally at odds with its ability to keep body movements under check at ridiculously high speeds. No matter what the surface, the car just copes.

So the Scooby's gone soft in its old age? Not likely. Subaru is now so adept at making four-wheel-drive cars handle neutrally – and in the new Impreza's case throttle-adjustable as well – that it's difficult to tell which wheels are actually doing the driving.

As before, traction out of any bend is superb. But put the power down mid-corner and the car just tightens its line and scoots off looking for the next bend. Back off in the same situation and the line tightens immediately. There's almost no understeer. And this hasn't come at the expense of a twitchy rear end either. Grip is stunning wet or dry, but a large lift will unstick the back tyres, courtesy of the raised rear roll centre.

And the biggest treat is still to come. This car's steering is quite brilliant, although it takes some time to get used to it. At first it feels too light and not quite sharp enough: under-sensitive even. Nor is it a particularly quick rack at 2.8 turns lock to lock. There's not that much resistance off the straight-ahead either, and in corners it doesn't load up very much.

But with time you learn to appreciate such subtleties and realise that this is one of the few systems that can play a sporting and cruising role equally well. Delicacy and accuracy are its two best traits. Combined with such a deeply talented chassis, it provides you with total confidence in the front end of the car and refuses to kick back even on the worst mid-corner bumps. And all Subaru did was modify the universal joint within the rack and mount it more rigidly. Extraordinary.

A near faultless performance

★★★★★

COMFORT, SAFETY AND EQUIPMENT

The Impreza WRX is a big improvement over the outgoing car inside, but it's still no Audi or BMW.

Subtle new curves shape a new dash and centre console arrangement that looks distinctly Japanese without being typically Subaru. The silver spray can has been put to work on the plastic surrounding the instruments and the hi-fi. It brings some relief to an otherwise dark space, though the cabin still looks plain compared with those of its more polished German rivals.

The instruments themselves aren't particularly attractive, but they're well placed and very visible. The driving position is spot-on, despite the fact that the steering wheel adjusts for rake only. The seats are typical Impreza: supportive and comfortable in equal measure, now height-adjustable too.

Cabin space is much better than before. There's 21mm extra width front and rear and a sizeable 30mm extra head room too. It may not be quite as roomy as a Ford Mondeo, but Subaru can rightly call the WRX a five-seater. The boot's a good size too, although the load height is taller than average.

Standard equipment is generous, but as ever with Subaru the list of options is short. As standard you get electric windows all round, CD player, air conditioning and those 17in alloys. But fancy interior materials, satellite navigation and a sunroof are all missing from the options list. The quality of the stereo is also far worse than in most rivals.

Safety standards are impressive. To start with there's the new bodyshell, which features more robust side impact bars and reinforced B-pillars and seat mounting points. Twin airbags at the front are standard, but side bags will cost extra and are only available with the optional sports seats (not fitted to the test car).

Practical and safe but still too bland

★★★★

ECONOMY

Shares in oil companies are rising, and they're set to rise further if the Impreza sells. Even if you refrain from using too many revs, fuel consumption remains a concern. We couldn't better 24.0mpg on our economy route, and charging across country this fell to a measly 13.4mpg. Worst of all, over the week-long period we drove the car it averaged only 18.7mpg, providing a disappointing real-world range of less than 250 miles from the 60-litre tank.

Simply too thirsty

★★

WRX has exceptional mix of rigidity and comfort on virtually any surface

THE AUTOCAR VERDICT

STAN PAPIOR

Almost the old car's performance, but now more refined. Great value, too

MARKET AND FINANCE

Looks could play a decisive role in the Impreza's fortunes. We're already hearing reports of customers cancelling orders because they can't get on with the shape. Nevertheless the order books are bulging.

There's nothing else to stand between the WRX and another display of residual values that will have rival manufacturers green with envy, assuming the car is as bulletproof as it was before. Subaru will restrict supply, so there will not be a glut of used cars, and there's no reason to expect anything other than a top performance from the dealers. It still comes with a full three-year/60,000-mile warranty.

Only doubt is over effect of looks on resale values

★★★★

WHAT IT COSTS

On-the-road price	£21,495
Total as tested	£21,495
EQUIPMENT	
(**bold** = options fitted to test car)	
Air conditioning	●
Traction control	-
Leather trim	-
Reach/rake-adjust steering column	-/●
Anti-lock brakes	●
Airbag driver/passenger	●/●
Satellite navigation	-
Alarm/immobiliser	●/●
RDS stereo/CD player	●/●
● standard - not available	
INSURANCE	
Group	17
Typical quote	£942
WARRANTY	
36 months/60,000 miles, 6 years rust	
SERVICING	
Major 15,000 miles	
Minor 7500 miles	
PART PRICES	
Oil filter	n/a
Air filter	n/a
Set of spark plugs	n/a
Brake pads f/r	n/a
Exhaust (excluding cat)	n/a
Door mirror	n/a
Tyre	£155
Windscreen	n/a
Headlamp unit	n/a
Front wing	n/a
Rear bumper	n/a

TESTER'S NOTES

Yet another new Bridgestone for us tyre geeks. This time it's called a Potenza RE 011 and, judging by the overall handling and ride of the Impreza, it's an absolute corker.

Why go to the expense of fitting a non-standard size hi-fi and then make it sound this limp? A head unit and speaker upgrade is what's needed. But then you've got a lovely hole in the dashboard. Not clever.

We'd like to say that there were a fair few folk who thought the Scooby rather attractive. But we'd be lying if we did. Everyone just grimaced and said: "Oh no, what have they done?" Sorry, but we agree.

It may not be anything special to look at in the daytime but at night the dash looks great. Everything has a spooky green glow to it. And in the dark you don't have to look at that silver dash surround.

It takes time to appreciate the new Impreza WRX's full range of qualities. At face value it appears to offer less than before. Less performance, less style. But the more miles we did in it, the more we realised Subaru has achieved just what it wanted, and has in the process built something even more special.

The key to its appeal is its drivetrain and chassis refinement. It is now an extremely comfortable long-distance machine. But it also offers an incredible depth of dynamic ability. It can't help but be a more satisfying ownership proposition.

No, it doesn't blow your socks off in a straight line like before, and yes it still drinks fuel at a worrying rate. But sample it over some choice A and B roads a few times, and then on a motorway, and you'll soon realise just how brilliant it is as a package.

And at £21,495 it's far cheaper than its German rivals, most of which aren't close to matching it as a driver's car.

If you can live with the looks, the WRX remains the definitive affordable performance saloon for the enthusiast.

Styling aside, a stunning achievement ★★★★★

Scooby does it

Subaru's new take on the cult Turbo might have divided fans, but it still shows the right spirit

Contender

Subaru Impreza 2.0 WRX turbo saloon £21,495

New model picks up where cult performance car and double *What Car?* award-winner left off

LEGENDS DON'T COME much bigger than the Subaru Impreza Turbo. It's King Arthur and the Loch Ness Monster rolled into one, the cult car of the last seven years, and current holder of the *What Car?* Junior Performance Car Award. So, to say that the new version, now called Impreza WRX, has much to live up to is like saying it's been quite wet this year.

On paper, Subaru makes all the right noises for the new car: better handling, equipment levels and build quality. So it's perhaps surprising that the first thing most people notice is the styling, particularly the nose.

Subaru says the oval headlights are 'unashamedly distinctive' and, we won't argue with that. Whether that's good is hotly debated. The rest of the car, though, with its spoilers, scoops and bulges in all the right places, still looks the part.

And let's not forget that the old car was no oil painting, yet it never struggled to find buyers. It won its converts with the way it drove, so that's the most crucial part of the new car, and first signs are good.

It hits 60mph half a second quicker than the old car, and shaves the same off the 30-70mph time through the gears. On the road, that gives even stronger, and safer, overtaking. Just drop down a cog or two, tweak the leather-clad Momo steering wheel to the right, give the throttle some grief and whatever was delaying you will soon disappear in the mirrors, all to the wonderful burble from the flat-four engine.

Perhaps more importantly, the brakes are more than a match. With an anti-lock system that knows exactly when to cut in, the brakes stop the car from any speed quickly and without fuss. Even on a damp test track, the stopping distances were up with the very best we've seen. On the road, it just adds to the driver's feeling of confidence; the pedal gives plenty of feel and a strong, progressive stop.

It's not all good news, though, because what's been gained in outright performance has been lost in flexibility. Thanks to higher overall gearing, and despite (a little) more pulling power at lower revs, the new car is slower in the gears. Compared to the old five-door (*What Car?*, February 2000), the new WRX is almost 1.5 seconds slower from 30-50mph in fourth gear and just under 2.0 seconds slower from 50-70mph in fifth.

The result is that the new car feels less urgent. Fifth gear soon becomes consigned to motorways and, even then, you occasionally need to drop down to fourth to pick up speed in the ebb and flow of traffic. In fact, it's all too easy to find yourself dawdling out of the power band and then there seems to be an age to wait before the engine wakes up above 3000rpm.

It's a similar story with the steering. While it's keen enough, with less than three turns lock to lock, it's lost some initial sharpness. Past that point, things sharpen up, but that, and improved refinement at speed, are further evidence of the car's slight softening.

Mind you, that's about as far as you can accuse the car of mellowing. With handling that's even sharper than the old model, it's one of those rare cars that flatters an average driver and delights a good one.

Through the bends, it's awesome. Even in torrential rain the WRX remains totally composed. Turn into a bend at speed and it remains perfectly balanced ready for you to squirt and aim for the next corner.

Grip levels are amazingly high and, even when pushing hard, you can adjust your line with a turn of the wheel or dab on the brakes. And, it just gets better the faster you go. The low-speed notchiness of the gearbox and slight sensitivity of the brake pedal give way to omplete smoothness once you're p and running. All the controls lend together beautifully.

Best of all, there's no sacrifice in ide. It's firm, of course, but never ncomfortable. Only the very worst imps and bumps thump into the abin and body control is never nything other than superb.

The biggest advances come in ther areas, though, notably build uality. Just one look at the dash is nough to see the improvements in s substance and clarity. Drivers an also choose between sports and lly-style bucket seats, although e latter means no side airbags.

The driving position is good, too, ith height adjustment on the river's seat and plenty of head and groom. If there is a gripe, it's that e steering wheel has height but ot reach adjustment. Otherwise, s good news all round. There's od space in the front and room in e back for a couple of adults with three rear passengers getting a ree-point seatbelt. There's a cent-sized boot, too.

As if that wasn't enough, the WRX so has much improved standard uipment. On top of (at least) two rbags and anti-lock brakes, xuries include air conditioning, mote central locking, a CD player d electric windows and mirrors.

Sadly, Subaru has yet to respond pressure from the UK importer to deadlocks. And, although the locks w have more shielding, that's a sappointment as our experts broke to the previous car easily in our curity Supertest (*What Car?,* ptember 2000). There is still a od immobilisation system, though.

All in all, there's some £2000 of tra kit over the old car for just a 00 price rise. With that in mind, no surprise that discounts are e. At the time of our test in early vember, very few importers or okers even quoted prices. The ly one that did, Dixon Motors, oted the list-price and that's a uation we don't expect to change ich. On the other hand, if the new car follows the old one, it will retain well over half of its value over three years and 36,000 miles.

Because the WRX is so new, it's a little early to predict running costs. But we do know that insurance is likely fall in to group 17 and fuel costs will be pretty high. Over some 400 fairly gentle miles during our test, mostly on motorways and without any sustained aggression, the Impreza couldn't top 26mpg. Make full use of the power, and you could struggle to match the urban figure of 19.9mpg, which is all the worse when it swallows £4-a-gallon super unleaded.

Verdict Worries about fuel costs are to ignore the Impreza's appeal. This is a driver's car above all else and, on that score, it's right on the line: quick with beautiful handling. Add improved build quality and better practicality, and – gripes about engine flexibility aside – it's hard to argue against it.

dling and grip are awesome; line on bends can be adjusted with turn of wheel or dab of brakes

Subaru Impreza WRX £21,495

Leg- and headroom good; choice of bucket or standard seats

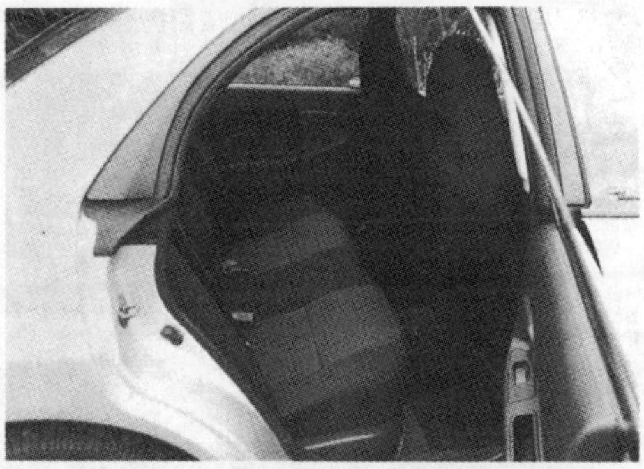

Rear seats pair of adults comfortably; three three-point belts

Build quality improvements are evident on dash; driving position good; height only steering adjustment

FOR

Superb handling; performance

AGAINST

Feels softer; less flexible to drive

VERDICT

Great update of cult classic, but lacks raw edge

●●●●○

MAKE	SUBARU
Model	Impreza 2.0 WRX saloon
On-the-road price	£21,495
Cost per mile	na
Fuel cost (10,000 miles)	£1475
Official mpg (urb/ex-urb/comb)	19.9/35.8/27.7
Service (3yr/36,000 miles)	na
Ins group/typical quote	17(est)/£942
Warranty (months/miles)	36/60,000
Engine (cc/type)	1994/4cyl horizontally opposed
Peak power (bhp/rpm)	215/5600
Peak torque (pulling power) (lb ft/rpm)	215/3600
0-60/30-70mph (sec)	5.7/5.7
Max speed (mph)	138
30-50mph in 3rd/4th/5th (sec)	4.5/7.9/15.3
50-70mph in 3rd/4th/5th (sec)	3.7/5.5/10.0
Braking 30-0mph/70-0mph (m)	9.7/39.9
Length/width/height (in)	173/76/57
Front headroom/legroom	excellent/excellent
Rear headroom/legroom	good/good
Boot size min/max (cu ft)	12.3
VERDICT	●●●●○

TEST DRIVE

TECH SPEC

SUBARU IMPREZA WRX

Engine: All horizontally-opposed four cylinder. Double overhead camshaft per bank, single turbo with intercooler. 215bhp at 5,600rpm, 215lb ft max torque at 3,600rpm.

Transmission: Permanent four wheel drive via five-speed synchromesh gearbox or four-speed auto. Torque split 50/50 (manual) 60 front/40 rear (auto).

Performance: top speed - 143mph. 0-60mph - 5.9sec. EC Urban fuel consumption - 19.9mpg.

Turbo boost: the new WRX features a bigger turbo to overcome the latest emission regs. But Subaru's trademark flat-four engine remains, lying short, low and in-line for optimum weight distribution

IMPREZA BIT OF KIT

MARK HALES has long been a fan of Scooby Doos - even before they were fashionable - so who better to test the all-new Subaru Impreza? He found it far from scrappy...

When I first started in this game, Subarus were part of a list which included big Fiats, Talbot Tagoras, big Lancias, Moskvitch and Polski. All of them cars which the magazines felt able to rubbish on a regular basis. Some of it was justified, but a lot was cobblers from cub reporters who picked the prevailing fashion then peddled it as wisdom. Subarus were definitely a victim of the latter. We had one arrive for road test and yes, it did look a bit odd - rather like two front halves welded together - but nobody in the office wanted to test it because they already knew it was rubbish. So it fell to me. I wrote at the time that while it might look strange, it drove extremely well and the combination of four wheel drive and flat four turbo engine was exciting and different. My enthusiasm served only to convince the experts that not only was the Subaru odd, but so was yours truly...

But it's all different now. The Impreza has been world rally champion in the hands of Scottish hero Colin McRae and is the rumbling electric blue roadburner of choice for the traffic light grand prix set. The number of evolutions and their body kits is heading towards double figures while the standard one is now a solid seller - one third of all Subarus made are sold in the UK. It's relatively cheap for what it offers and holds its price extremely well. Strange no longer... And yet they are. Subaru makes the only car in the

world powered by a watercooled flat-four engine, and the only saloon car range which features four-wheel drive as standard - there is no longer any two wheel drive option. But more significant is that you would never describe an Impreza or Legacy as slinky exercises in the art of styling. Purposeful, perhaps, but not exactly beautiful. In a time where image is all, people obviously buy Subarus for the engineering.

That engineering has recently received an update, as has the, ahem, styling. The shell is slightly bigger by a couple of inches in most directions, is stronger and stiffer - by a claimed 250 per cent so it must have been kinda floppy before - and offers better crash resistance. The exterior style borrows some of that rally car look and big arches are swaged into the rest to create a more aggressive presence which is set off by big oval headlights and a big chin moulding. To go with the stiffer floorpan and shell, suspension geometry has been refined to make the car more responsive although there are still struts at both ends, whereas the larger Legacy has a multi link rear. Subaru says this is because the strut has greater travel and is preferable in a more sporty car - which sounds a bit like accountant's spin to me... The top of the range model is now called WRX and this boasts some fashionable forged aluminium track control arms and a stiffer subframe as well as a bigger turbo to overcome the additional catalyst required for the latest emission regs. Power stays the same but mid-range torque goes up and there's a honky racer's scoop on the bonnet to feed air to the intercooler which sits on top of the engine and makes use of the extra bonnet space liberated by the horizontally opposed cylinders.

Clock watching: the Impreza WRX gets from 0-60 in less than six seconds while top speed is 143mph

There are a lot of little detail changes too, like wider track, revised roll centres, stiffer suspension bushing, different springs and dampers, revised pedal ratios for better brake feel, revised speed sensitive power steering for more feedback and so on. A list that does seem to focus on the intimate side of the driving experience and which supports what Subaru keep saying - that their baby is intended for people who appreciate driving.

Crash protection is better and the interiors offer more space, but remaining exactly the same is the basic layout of the flat-four engine and the simple permanent four-wheel drive system which have been Subaru sacred cows since the 1960s. Subaru's rationale for this has always been the pursuit of safety combined with driving pleasure; all-drive spreads the power amongst more wheels which improves traction, while a flat four engine is shorter and lower which means the major mass is lower and further back. That gives better weight distribution, added to which the engine runs in-line rather than across the car which means the all-drive system can be simpler, lighter and more compact. According to the engineers it adds about 100lbs which they reckon is half as much as anybody else's.

The Japanese engineers along for the launch were also refreshingly candid about a flat four's less good points. There are two cylinder heads instead of one which is normal for a four cylinder engine, and the exhaust they say is a nightmare - especially with a turbo - and they are still working on refinement.

But is this one any better than the last? I found a WRX which sported an interior which was unremittingly black - relieved only by a bit of essential silver on the dash - and with seats that looked plastic but weren't, and which turned out to be extremely supportive and offered a driving position which was close to perfect. Then the raggedy beat of the flat-four engine is still so different from anything else, you can confuse the rattle of the gearbox at low revs with harshness; it's not, but sounds like it.

Drab dash: despite the silver detailing Hales found the interior design of the WRX a little too black

That said, it's better than it was thanks to new gearboxes and drivetrains and a new crankshaft damper but it seems a strange irony that a layout which is supposed to cancel out vibration always seems to rattle at low revs. That's something else the engineers say that they are working on...

Once properly under way and on roads made treacherous with squalls of rain, the Impreza is, as ever, supremely sure footed - almost worryingly so. You can press on with such abandon that you get little sense of how fast you are going - which should be taken as a compliment - and it is always entertaining without demanding any great skill at the wheel. Meanwhile, the ride is very good; Subaru make great play of their long-travel suspension, but they have included excellent body control without rattling the occupants.

It is also good and fast. The book says the turbo WRX gets to 60mph in under six seconds but this doesn't describe the way you just leave it in third and gobble up twisty roads dotted with traffic. Common sense and a desire to see the next birthday was the necessary limiting factor here but included in the day's events was a visit to an even more treacherously greasy Knockhill circuit. Maybe a surface that has been hammered by Scottish hardmen for decades has acquired a sheen which is slipperier than most, but then such conditions should be ideal to prove that all-drive is a good thing.

Sure enough, booting it good and early in the turn only made the front run wide - even newly fattened turbo torque could not unstick the wheels - but you could still get yourself in trouble. The regular Knockhill lads who instruct there all the time had done a few laps by then and their advice was always the same: It's great as long as you keep your foot in it... And make no mistake, if you did back off in the middle the tail was way out of line in a trice. There's no doubt

SUBARU IMPREZA RANGE
On the road price/availability; Impreza 2.0WRX manual saloon £21,495, Sports Wagon £21,995. 2.0GX £15,750, Wagon, £16,250, auto saloon £16,750. 1.6TS manual saloon £13,950. Available end November.

Subaru blue: the colour's nice but no Subaru has ever been seen as a work of art when it comes to looks. Functionality is the thing - and fans would have it no other way

that all-drive is the ultimate aid to traction but it can only influence the handling when you accelerate. Even so, it was nice to find out and to be fair you'd be unlikely to go that fast on a public road - at least not one that felt as slippery as Knockhill did. And besides, if Subaru tweaked out that last bit of entertainment, the car probably wouldn't be so nice to drive.

There is an estate variant and the option of two-litre non turbo and 1.6 litre engines - both flat fours naturally - which after the WRX seemed to have so little power relative to the amount of traction and grip that you wondered whether all-drive was a touch over the top. On the other hand, since Subaru don't charge any premium for the technology, it represents added value that might come in handy one day. And this is what the Impreza is all about. The volume versions may not look very special but once you get in and drive you find a deal of extra capability which has been made so much a part of the design and performance that you almost don't notice it. It is also extremely good value. At £21,495, the turbo two-litre WRX is £7,400 cheaper than an Audi A4 2.8 quattro, and is likely to maintain its worth. A car dealer chum reckoned that an Impreza which cost £20,000 when new is still worth £15,000 two and a half years later. He was also able to supply a bit of dealer folklore. "People who buy them," he said, "are definitely into their cars. And although they're not going to buy anything else, they seem to spend more time thinking about it than any of my other customers." ■

NEWCOMERS

SUBARU IMPREZA WRX STi

Hard and fast

Subaru tweaker STi has worked its magic on the new Impreza. It's not yet definite when it arrives here, but it's a better, sharper package

AROUND THE NÜRBURGRING, the new Subaru Impreza WRX STi is said to be the business. It's the rocket power of the 280bhp specially tuned engine that does it. That and the fine, new six-speed 'box Subaru's come up with. Stiffer suspension, too, plus those gumball tyres and meatier Brembo brakes. Put it all together with the customary full-time 4WD and you have, for now at least, the ultimate WRX of the new era, a car that lists in Japan for the equivalent of £19,000. Outrageous.

Nürburgring or not, you quickly get the measure of the STi and what it's all about. In real life, on normal roads, it's like the new Japan-spec 250 bhp WRX (on which it's based), only more so. As in quicker, with more grip, more exotic and, remarkably, just as refined. Against the new UK Rex (215 bhp) the gulf widens still further. But that's life.

For those just tuning in, STi is Subaru Technica International, the factory-backed outfit outside Tokyo that develops all the tweaky tuning bits for Imprezas, Legacys and Foresters. Of course, STi did a whole run of cult, go-faster Mk1 Imprezas that with Colin McRae *et al* helped seal the car's fame. Now it's a first run at the new-shaped car and in Japan you get to choose either the four-door saloon body STi or five-door sportswagon, all for around the same money.

First off, looks. Those expecting something mad will be sadly disappointed. Aside from those tasty gold alloys and tiny STi-badged spot-light covers, visually, it's business as usual. And while we're still learning to love the new Impreza shape, the WRC blue (one of four colours available) does help things along. As for the optional HID lights on Japanese Imprezas, erm...

In the great scheme of things, STis are homologation specials, built for the semi-pro race/rally driver and serial road-going WRX'ers. For those who really will use their STi in competition, Subaru does the type RA, a lighter, bare-spec special with faster-ratio steering and more rear-biased 4WD with lockable centre diff. Saves 50kg at the kerb, too.

Standard STi or type RA, the burbling 1994cc flat-four turbo engine's the same. Block, heads, pistons and con-rods are all tougher. There's a bigger turbo and intercooler and, like the 250bhp engine, it gets the AVCS variable-cam tech that UK cars don't. It's an engine you can use harder and rev higher, and that's the point.

Though there's a 8000rpm redline, a guardian-angel buzzer and bleeper come on at 7500rpm, just to make sure. The engine's fabulously smooth and revvy but has a narrow and quite ferocious power band. Between three and seven is where the action is, and running hard in that mid-range groove, the STi's almost insanely fast, the power

The gearbox gains another ratio and an improved shift feel

Better rubber and stiffer springs make STi more demanding

Some STi badges and funky blue-black seats to differentiate

arriving in one big wallop and you'd better be ready.

By the same token, it's no hardship to punt around town. Traffic queues betray no temperament and the light, precise, machined shift feel of the new close-ratio six-speed 'box is just great. For motorway cruising, the extra top cog also reduces engine speed, noise and improves fuel consumption, so it's cool all round.

Subaru claims 5.1secs for the 0-6 sprint, with 0-400 metres taking 13.2 secs. The Japanese brochure talks about a 149mph top end.

The goalposts have moved with the new WRX, of course. Less 'edge', less clunky now, and while the STi gets 30 percent stiffer springs, a front-strut tower bar and higher grade road rubber, it hasn't gone all hardcore and tough. For one thing, ride quality, while firm, is no boneshaker. In fact, in this Premier Division, the STi rides amazingly well. Bad news, perhaps for some of you out there, but there you go.

The STi connection means less body lean in corners and tighter turn-in. But still nothing like the taut, crazed, full-on feel of a Mk1 WRX or a Lancer Evo, for that matter. Though immensely quick, the new WRX isolates you more from the action and with the STi, the manic turbo surge can also upset handling, so feeding in the power on approach to a tight apex calls for just a bit of caution. Bottom line: compared to lesser WRXs, A-road driving is more demanding.

At the other end, the STi is well tied down, yet happy enough to push sideways on demand or, should you suddenly lift-off, mid-bend. Big 17-inch Brembo vented discs have all the muscle spec but feel somewhat dead unless you really work them hard. Steering is quick, accurate, well-weighted, making the STi easy to place but could do with more feedback. As for the STi turning circle, it's not small.

Inside? Just like the new UK Rex, save for different blue-black seat trim and assorted STi badges. Speedo and tacho also swap places. A set of aluminium 'anti-slip' pedals and footrest arrive for duty and the legendary intercooler water spray button makes a comeback.

So is this all academic? Subaru hopes for Whole Vehicle Approval for the 280bhp car by the end of the year. Single Vehicle Approval would be a logical resort. The exchange rate means a P1-ish £30-32k price range for an official car, or grey imports. If money is no object, call your local grey merchant for a 12-week delivery period.

PETER NUNN

Rating	★★★★★
Price	£32,000 (est)
Engine	1994cc 16V four, 280bhp, 275lb ft
Performance	149mph, 0-60mph 5.1secs, 29.4mpg
On sale in UK	TBA

STi of the storm

DRIVE
Subaru Impreza STi

With scorching performance and awesome traction, is the new STi the best Impreza ever? By Stephen Sutcliffe

STi's interior feels much plusher than the regular Impreza WRX, with alloy-effect plastics and blue cloth; new sports seats are comfortable, supportive and stylish

SUBARU IMPREZA WRX STi
PRICE £31,995
ON SALE Now

OFFICIALLY, THE 280bhp, 155mph Subaru Impreza STi you see here does not exist yet – at least not in the UK. Subaru's official importer wants to make us wait 18 months or more before we can buy one here.

Which is a great pity because, having just spent the day howling around the Cotswolds in one, I can tell you: it's one hell of a car.

No matter. If you can't wait until 2002 to get your hands on one (and if you like this sort of car, I strongly suggest you don't wait), there is an alternative.

Step one, make sure you have £31,995 in your bank account. Step two, get a map and find Malmesbury on the map (it's in Wiltshire, about 10 miles south of Cirencester, just off the M4). Step three, get yourself along to David Hendry Sports Cars (tel 01666 824369) and ask them nicely if you can have a go in one of their demonstrator STis. Step four, get ready to make one of the easiest purchasing decisions of your life.

Why? Because if you're into this type of car you will, I guarantee, not be able to resist writing out a cheque there and then. So make sure you tell the missus where you're going before you set off.

Ah, but surely you don't get a warranty if you buy through a grey importer like Hendry, I hear you say. And the speedo probably reads in kilometres per hour while the radio tries to pick up Hiroshima FM.

Nope. Hendry will make all the alterations for you and throw in a three-year warranty and Thatcham category one alarm. Which means what you're looking at is a rather delicious piece of motoring future, right here and now, courtesy of some quick and clever thinking on the part of the Hendry family.

If you're somehow not familiar with the Impreza Turbo story, it

Brembo calipers match gold 17in alloys; ventilated front discs measure 294mm

When speedo is changed to mph it will have to cope with top speed of 155mph

Straight-line performance is awesome. But what's even more impressive is the excellence of the STi's steering – especially as the ride is better than the Evo VI's

goes a bit like this. For many years there was a company called Subaru, which produced weird and not very wonderful machines that few people cared about, cars like the XT coupé and the Justy.

Then came the Impreza, since when Subaru has been regarded as a kind of thinking man's alternative to BMW with four-wheel drive and a slice of left-foot braking thrown in for free.

Well, just as BMW has its M division to turn humble Threes and Fives into hot rods, so Subaru has STi – Subaru Technica International. And when the STi guys gets their teeth into something, the results are usually spectacular – and never more so than they are here.

Outwardly there are myriad styling tweaks that differentiate the STi Impreza from the regular WRX, and they create a much more dramatic effect in the metal. The sills are colour coded, the air intake on the bonnet is bigger, the wheels are gold and only partially conceal huge great Brembo brake calipers which are also sprayed gold.

The rear windows on this particular example are also blacked out, though officially that's not legal in the UK so the factory cars won't be equipped with these. Pity this, because although it sounds tarty, the Marmite glass lends the car a lot of extra presence on the road.

Specification wise, the STi Scooby is a technophile's dream come true. Lift the bonnet and you're greeted by the most amazing sight of hundreds of different coloured leads and hoses all beautifully integrated in and around the 1994cc turbocharged boxer engine. There's also a huge air-cooler with STi inscribed in among the fins in bright red ink.

Subaru claims as much as 80 per cent of this engine is new, but the bits that count are the new variable valve timing system, the bigger air-to-air intercooler and the forged pistons that help it rev beyond 8000rpm. Power is officially quoted at 280bhp, but only because that's the limit imposed by the Japanese legislators. In reality Hendry reckons it has about 310bhp at 6400rpm, as well as a lot more torque than the old STi VI engine, the official figure being 275lb ft at 4000rpm.

On their own, these improvements would be enough to ensure the STi had blistering performance compared to the previous-generation WRX, despite putting on a few extra kilos largely in the name of safety. But when mated to a new close-ratio six-speed gearbox they're said to make even more of a difference. Lag, by all accounts, is now virtually non-existent on the road – a claim that, on paper, is hard to believe.

The chassis is fundamentally the same as the ordinary WRX's with one or two STi bits thrown in for good measure. There's a slightly quicker steering rack, uprated bushes and beefier suspension arms front and rear, and bigger Brembo brakes. But the most significant difference is the fitment of a limited-slip differential at the front, to match the one at the back; a vital piece of kit denied drivers of the regular WRX. This isn't fitted to all STis: in Japan the car comes in three trim levels ranging from KL (no front diff), via EL (front diff but no gas discharge headlights or blacked-out rear windows) to GL, which gets the full complement of hardcore hairiness. It's almost certain that a bastardised version of the GL is what we'll get in 2002. And naturally it's a GL that Hendry has let us loose in.

Snuggling down into the high sided and rather smart looking STi-logoed seat, it's obvious how much plusher the cabin is compared to the regular Impreza WRX. Slivers of alloy-look plastic surround the gearlever, stereo and heater controls, and the instruments are chrome-rimmed too. There's even a button to ➧

six-speed manual 'box a delight to use

adjust the point at which the change-up buzzer sounds, while the water spray switches of the previous STi VI are thankfully still present. While being able to blow a fine spray of water across the face of the intercooler may not be the most vital piece of standard equipment to make use of on your way down to Tesco's, it's all part of Subaru's way of preparing you for what's to come.

Except, of course, nothing can prepare you for what this car is capable of on a wet and windy B-road. Not even a Mitsubishi Evo VI. Yep, you read that correctly: the STi is at last the car that puts the Evo in its place. And I know this from experience, having put the two back to back for the day, more of which you can read about next week. Although judging from our first drive of the Evo VII (see p28) it may not stay that way for long.

In-your-face styling includes gold alloys, bigger air intake and tinted rear windows

Bigger STi-badged intercooler helps 2.0-litre boxer engine unleash up to 310bhp

To begin with, the only aspects that appear notably different from the regular WRX are the firmer ride (it's hard but not crashy), the slightly heavier and quicker steering, the meatier brakes and the super-short throw gearchange. But the first time you put your foot down and hold it there for more than a second, one very big difference hits you: unlike the regular WRX, the STi goes. Really goes. Just guessing, I'd say it'd do 60mph somewhere in the mid fours and 100mph in 10-11sec. So in other words it's as quick as, if not quicker than, an Evo.

Subaru's claims that the STi has less lag are true up to a point but there's still a delay between opening the throttle and the meat of the acceleration actually starting – about a second by my reckoning. But once the mayhem starts it just keeps on coming, and all you do is pick off another gear before the 8100rpm limiter intrudes to sustain the outrageous momentum.

Get the STi moving above 50mph on A and B-roads and fourth, fifth and sixth are the only gears you need to use. Third is just a toy to bring out on special occasions while second is purely for showing off.

Yet what makes the STi so eye-wateringly capable across country, and so unique among its peers, is not its engine or its straight-line performance. It's the chassis, and more specifically its steering precision and basic suspension control. Thanks to all the electronic trickery contained within the various diffs front and rear it always has massive traction, wet or dry.

But unlike the Evo it also has an uncanny ability to soak up intrusions and simply glide over rough ground seemingly unhindered by scars and ruts. It also suffers from a lot less steering fidget than an Evo and is simply more composed for more of the time over rough surfaces, which is an absolutely crucial element considering the state of most of our roads.

No, it's maybe not as raw-knuckle exciting over a 10 minute balls-out blast. But once the red mist fades and you're left with the realities of everyday motoring, it's convincingly the better car. I think I'd even rate it above the forthcoming new M3 in most respects, simply because you can use more of its performance without attracting so much attention to yourself while doing so. In an era of cameras and anti-speed campaigning, such qualities are increasingly relevant.

The only question mark over the Impreza STi is when and where to buy one. Either you can wait two years for the official versions to filter through to Britain, or you can order one now via an outfit like Hendry. Doesn't sound like much of a dilemma to me. ⭘

FACTFILE

SUBARU IMPREZA WRX STi

HOW MUCH?	
Price	£31,995
On sale in UK	Now
HOW FAST?	
0-60mph	4.6sec (est)
0-100mph	11.0sec (est)
Top speed	155mph (est)
HOW THIRSTY?	
Best/worst	24.4/14.9mpg
HOW BIG?	
Length	4405mm
Width	1990mm
Height	1440mm
Wheelbase	2525mm
Weight	1400kg
Fuel tank	60 litres

ENGINE

Layout 4 cyls horizontally opposed, 1994cc
Max power 280bhp at 6400rpm (see text)
Max torque 275lb ft at 4000rpm
Specific output 140bhp per litre
Power to weight 200bhp per tonne
Torque to weight 196lb ft per tonne
Installation Front, longitudinal, four-wheel drive
Bore/stroke 92.0/75.0mm
Made of Cast alloy heads and block
Compression ratio 8.0:1
Valvegear 4 per cyl, dohc per bank
Ignition and fuel STi electronic ignition, IHI turbocharger, intercooler

GEARBOX

Type 6-speed manual by STi
Ratios/mph per 1000rpm
1st 3.64/5.2 **2nd** 2.37/8.0 **3rd** 1.76/10.8
4th 1.35/14.1 **5th** 1.06/18.0
6th 0.84/22.7 **Final drive** 3.90

SUSPENSION

Front MacPherson struts, coil springs, anti-roll bar
Rear MacPherson struts, coil springs, trailing arms, anti-roll bar

STEERING

Type Rack and pinion, power assisted
Lock to lock 2.7

BRAKES

Front 294mm ventilated discs
Rear 266mm ventilated discs
Anti-lock Standard

WHEELS AND TYRES

Size 7.0Jx17in
Made of Cast alloy
Tyres 225/45 ZR17 Bridgestone RE040

All figures are manufacturer's claims, except those relating to performance

VERDICT

The definitive evolution of a definitive car. The Evo VII will have to be truly special to win the crown back for Mitsubishi.